RUSSIAN FEDERATION
West Siberian Plain
Ural Mountains
KAZAKHSTAN
The Steppes
MONGOLIA
CHINA
UZBEKISTAN
TURKMENISTAN
KYRGYZSTAN
TAJIKISTAN
AFGHANISTAN
IRAN
IRAQ
SYRIA
TURKEY
GEORGIA
ARMENIA
AZERBAIJAN
SAUDI ARABIA
Arabian Peninsula
YEMEN
OMAN
QATAR
BAHRAIN
KUWAIT
LEBANON
ISRAEL
JORDAN
U.A.E.
Yamal Peninsula
Gyda
Putorana Plateau
Mezhdusharskiy I.
Krasino
Kolguyev Island
Vaygach Island
Murmansk
L. Imandra
Kola Peninsula
Nar'yan Mar
ARCTIC CIRCLE
Vorkuta
Salekhard
Pechora
Mezen
1895
Arkhangel'sk
L. Vyg
Petrozavodsk
Lake Onega
Northern Dvina
Syktyvkar
Ob
Khanty-Mansiysk
Siberian
Lower Tunguska
Tura
Gravelly Tunguska
Yenisey
Lake Beloye
Sukhona
etersburg
Rybinsk Reservoir
Vologda
Vyatka
Kudymkar
Reservoir of the Kama
Irtysh
ovgorod
Yaroslavl'
Kostroma
Ivanovo
Tver
Nizhniy Reservoir
Nizhniy Novgorod
Yoshkar-Ola
Kama
Perm'
Izhevsk
Tyumen
Yekaterinburg
Tomsk
Krasnoyarsk
Bratsk Reservoir
MOSCOW
Vladimir
Cheboksary
Kazan'
Kurgan
Omsk
Novosibirsk
Kemerovo
Ryazan'
Samara Reservoir
Chelyabinsk
Tobol
Petropavlovsk
L. Chany
Eastern Sayan Mts
Kaluga
Tula
Saransk
Simbirsk
Ufa
Ishim
Barnaul
Abakan
2922
Bryansk
Orel
Penza
Saratov Reservoir
Tol'yatti
Samara
Kustanay
Kokshetau
Lake Selety
Gorno-Altaysk
Western Sayan Mts
Lipetsk
Saratov
Volga
Orenburg
ASTANA
Kyzyl
3492
Hövsgöl Nuur
Kursk
Tambov
Ural'sk
Ural
Lake Tengiz
Semipalatinsk
Ust-Kamenogorsk
3121
Uvs Nuur
Belgorod
Kharkiv
Voronezh
Aktyubinsk
Karaganda
4506
Belukha
4374
Hyargas
Dniepr
Don
Volgograd
Lake Zaysan
ropetrovsk
Donets'k
Caspian Depression
Atyrau
Dzhezkazgan
Rostov-na-Donu
Elista
Astrakhan'
Lake Balkhash
Alakol
Crimea
Krasnodar
Stavropol
Betpaqdala
Taldykorgan
Sevastopol
Maykop
Cherkessk
Nal'chik
Grozny
Aral Sea
Kyzyl-Orda
Almaty (Alma-Ata)
Shihezi
Urumqi
5642
Elbrus
Ust-Urt Plateau
Kyzyl-Kum
7439
Pik Pobedy
Yining
Caucasus
Vladikavkaz
Makhachkala
Kara-Bogaz Bay
Nukus
Chimkent
BISHKEK
L. Ysyk
Tian Shan
Sinop
TBILISI
Urgench
TASHKENT
Namangan
Andizan
ANKARA
YEREVAN
BAKU
Osh
Kashi
Samarkand
Khujand
Chardzhev
Turkmenbashi
Kara-Kum
Amu Darya
7719
Kongur Shan
Mt Ararat
5137
4810
Kayseri
L. Van
L. Urmia
Tabriz
ASHGABAT
Mary
DUSHANBE
Qilian Shan
Qinghai Hu
Xining
Taurus Mts
Gaziantep
KURDISTAN
Gorgan
Mazar-e Sharif
8611
K2
7690
Mersin
Adana
Aleppo
Al Mawsil
Tigris
Euphrates
Mt Damavand
5671
TEHRAN
Mashhad
KABUL
NICOSIA
3088
BEIRUT
DAMASCUS
Kirkuk
Kermanshah
Qom
Salt Desert
Herat
5143
Syrian Desert
BAGHDAD
Mesopotamia
Tel Aviv-Jaffa
AMMAN
JERUSALEM
Dead Sea
Razzaza
Ahvaz
Isfahan
Lut Desert
Kandahar
Suez Canal
Sinai
2637
2580
Basra
Abadan
KUWAIT
Kerman
Shiraz
Zahedan
Great Nafud
Hejaz
Buraydah
Ad Dammam
Bandar-e Abbas
Qeshm
MANAMA
DOHA
Dubai
Medina
RIYADH
ABU DHABI
TROPIC OF CANCER
Harad
MUSCAT
Jeddah
Mecca
Masira
Rub al Khali
Abha
Salalah
SANAA
3760
Aden
1503
Socotra
YEMEN

Terralog

Patrick David
Gernot Vogel

Venomous Snakes of Europe, Northern, Central, and Western Asia

Giftschlangen Europas, Nord-, Zentral- und Westasiens

Edition Chimaira

A.C.S. GLASER GERMANY

David, Patrick & Vogel, Gernot:
TERRALOG Vol. 16:
Venomous Snakes of Europe, Northern, Central and and Western Asia/Giftschlangen Europas, Nord-, Zentral- und Westasiens
Frankfurt am Main 2010:
Edition Chimaira

ISBN 978-3-89973-366-2

Front cover/Umschlagvorderseite:
large photo/großes Foto:
Vipera orlovi. M. Dobiey
smaller photo to the right/
kleineres Foto rechts:
Atractaspis engaddensis. D. Modrý
smaller photo to the left/
kleines Foto links:
Echis omanensis. D. Donaire

Full page photo opposite to symbol explanations/ ganzseitiges Foto gegenüber den Symbolerklärungen: *Vipera ammodytes ammodytes*. D. Bohle

Heddernheimer Landstr. 20
60439 Frankfurt am Main
Germany
Phone: +49-69-497223
Fax: +49-69-497826
E-Mail: frogbook@aol.com
Internet: www.chimaira.de

Publisher/Verleger: Andreas S. Brahm
Layout/Satz: Sibylle Manthey
Photo Editing/Bildbearbeitung: Ulrich Manthey
Printing/Druck: DCM, Meckenheim, Germany
Binding/Bindung: Freitag, Kassel, Germany

ISBN 978-3-89973-366-2

Contents/Inhalt

Introduction

This volume is the third in the series on Venomous Snakes of the World. Previous volumes covered Tropical Asia (Vogel 2006) and Africa (Dobiey & Vogel 2007). The present volume is dedicated to the venomous snakes of Europe, Asia Minor, the region of the Caucasus, the Near and Middle East, the Federation of Russia (including the Far East), countries of Central Asia, and Mongolia (see maps on the inner front and back covers). Species inhabiting the Sinai Peninsula, excluded from the volume on Africa, are treated in the present volume.

How to use this book

This volume contains all land species occurring in the region delimited above and belonging to the families Elapidae Boie, 1827, and Viperidae Oppel, 1811. We also include two species of the genus *Atractaspis* belonging to the recently resurrected family Lamprophiidae Fitzinger, 1843 (see Vidal et al. 2007, 2008). Although various sea snakes are known to occur on coasts of southeastern Russia (Ananjeva et al. 2004, 2006), these are not considered here.

We have opted to not include in the present volume *Rhabdophis tigrinus* (Boie, 1826), which is present in extreme southeastern Russia (Ananjeva et al. 2004, 2006). Despite the fact that this species is known to have caused human fatalities in Japan (see David & Ineich 1999), our definition of venomous snakes is limited to genera with a fully functional venom delivery system. By default, this also excludes from the present volume all opisthoglyphous colubrids.

A map outlines the known distribution for each species. Maps like these should be taken as an indication of the possible presence of the respective taxon within the outlined area; natural habitats and elevations need to be taken into consideration for more precise distribution maps.

It was our intention to present taxonomical positions as up to date as possible, which does not necessarily mean they reflect the most conservative and/or consensual concepts. Published revisions were taken into consideration up to December 1st, 2009. Common English / German names were adopted from various sources, for example from Mallow et al. (2003), Joger & Stümpel (2005) and De Smedt (2006). Names introduced by ourselves are marked with an asterisk (*).

We would like to stress on the fact that keeping venomous snakes in captivity is by nature hazardous. This book should not be considered, in any way, to be an invitation or encouragement to keeping and breeding venomous snakes in captivity. Venomous snakes should be kept only by keepers with the necessary expertise and suitable facilities and in full compliance with national and regional regulations.

Some guidelines for the captive husbandry of each taxon are provided at the end of this book. Readers should regard the given terrarium dimensions as minimum requirements. Many of these suggested dimensions were calculated according to the guidelines for the captive keeping of wild animals issued by the German Ministry of Food, Agriculture and Forestry (BML).

Einleitung

Der vorliegende Band ist der dritte in der Reihe über die Giftschlangen der Welt. Vorangegangene Bände decken das Tropische Asien (Vogel 2006) bzw. Afrika (Dobiey & Vogel 2007) ab. Der gegenwärtige Teil ist den Giftschlangen Europas, Kleinasiens, den Ländern des Kaukasus, des Nahen und Mittleren Ostens, Russlands (einschließlich des Fernen Ostens) Zentralasiens und der Mongolei gewidmet (vgl. die Karten auf den Innenseiten des Einbands). Auch die im Afrika-Band nicht berücksichtigten Arten der Sinai-Halbinsel werden hier abgehandelt.

Benutzerhinweise

Dieser Band umfasst sämtliche landlebenden Arten, welche in der oben umrissenen Region vorkommen und den Familien Elapidae Boie, 1827, bzw. Viperidae Oppel, 1811, angehören. Eingeschlossen sind weiterhin zwei Arten aus der Gattung *Atractaspis*, die zu der kürzlich revalidierten Familie Lamprophiidae Fitzinger, 1843 (siehe Vidal et al. 2007, 2008) gehören. Obwohl verschiedene Seeschlangen als an den Küsten Südost-Russlands vorkommend bekannt sind (Ananjeva et al. 2004, 2006), werden diese hier nicht berücksichtigt.

Rhabdophis tigrinus (Boie, 1826), die im äußersten Südosten von Russland vorkommt (Ananjeva et al. 2004, 2006), wird ebenfalls nicht behandelt. Wenngleich die Art als Verursacher von tödlichen Bissunfällen beim Menschen in Japan bekannt ist (siehe David & Ineich 1999), ist unsere Definition von Giftschlangen auf jene Gattungen beschränkt, die über ein voll funktionsfähiges Giftinjektionssystem verfügen. Demzufolge sind auch alle opisthoglyphen Colubriden vom vorliegenden Band ausgeschlossen.

Das bekannte Verbreitungsgebiet jeder Art wird anhand einer Karte dargestellt.

Es war unser Ziel, die Stellung jedes Taxons nach den neuesten Erkenntnissen wiederzugeben. Dies bedeutet jedoch nicht zwangsläufig, dass es sich dabei in jedem Fall um das konservativste und/oder ein allgemein anerkanntes Konzept handelt. Veröffentlichte Revisionen wurden bis zum Stichtag 1. Dezember 2009 berücksichtigt. Die englischen und deutschen Trivialnamen wurden von diversen Quellen übernommen, darunter vor allem von Mallow et al. (2003), Joger & Stümpel (2005) und De Smedt (2006). Von uns hier erstmals eingeführte Trivialnamen sind mit einem Stern (*) gekennzeichnet.

Wir möchten betonen, dass die Haltung von Giftschlangen im Terrarium von Natur aus risikoträchtig ist. Dieses Buch soll daher auch in keiner Weise als Einladung oder Ermutigung zur Haltung und Zucht von Giftschlangen in Gefangenschaft verstanden werden. Giftschlangen gehören ausschließlich in die Hände von Haltern mit der nötigen Erfahrung und den erforderlichen Einrichtungen und sollten nur unter strikter Beachtung der nationalen und regionalen gesetzlichen Vorschriften gehalten werden.

Am Ende dieses Buches finden sich einige Richtlinien zur Terrarienhaltung für sämtliche abgehandelten Taxa. Der Leser sollte hier die angegebenen Terrarienabmessungen als Mindestmaße verstehen. Viele der Abmessungen wurden anhand der vom deutschen Bundesministerium für Ernährung, Landwirtschaft und Forsten herausgegebenen Mindestanforderungen für die Haltung von Reptilien errechnet.

Taxonomy

The basis of our classification at generic level and below is formed by the work of David & Ineich (1999) and the various contributions in Joger & Stümpel (2005), as modified by subsequent revisions, and of Stümpel & Joger (2009). These recent modifications are specified below in the respective familial and generic accounts. Suprageneric classification is based on Vidal & Hedges (2002) and Vidal et al. (2007, 2008). In agreement with the current classifications, the family Viperidae is divided into the subfamilies, Viperinae and Crotalinae.

ELAPIDAE Boie, 1827

Genus *Bungarus* Daudin, 1803

Pending the publication of the revision announced in Vogel (2006), we accept only two species for the region covered by this volume. *Bungarus caeruleus* (Schneider, 1801) does not call for comments. Khan (1985) revised this species and described *Bungarus sindanus razai* Khan, 1985, the validity of which is questionable (Vogel 2006). This species was recently discussed by Kuch (2004).

Genus *Naja* Laurenti, 1768

This genus has been revised by Wallach et al. (2009), who divided it into four subgenera. Long considered a subspecies of *Naja naja*, *Naja oxiana* (Eichwald, 1831) was granted full specific rank by Wüster & Thorpe (1991, 1992a-b) and Wüster (1996); see also Ananjeva et al. (2004, 2006). The former subspecies *Naja haje arabica* Scortecci, 1932, which belongs to the predominantly African complex of *Naja haje* (Linnaeus, 1758), has been raised to full species status by Trape et al. (2009).

Genus *Walterinnesia* Lataste, 1887

This genus, largely endemic to the Near and Middle East, has been revised by Nilson & Rastegar-Pouyani (2007) who resurrected *Walterinnesia morgani* (Mocquard, 1905) for eastern populations.

LAMPROPHIIDAE Fitzinger, 1843

Genus *Atractaspis* Smith, 1849

The intricate taxonomy of this largely African genus was discussed by Dobiey & Vogel (2007). Only two species are known from the region covered in this book, both of which are is still controversial as to their systematic status. We follow Vidal et al. (2007, 2008) in placing the genus *Atractaspis* in the subfamily Atractaspidinae of the family Lamprophiidae.

Atractaspis microlepidota Günther, 1866, has for a long time been included in the fauna of the Near East (for a summary see David & Ineich 1999). Trape et al. (2006) revised the group of *Atractaspis microlepidota* Günther, 1866, with *A. microlepidota* being restricted to Senegal, Gambia and Mauritania in the process. The authors resurrected *Atractaspis watsoni* Boulenger, 1908, for populations of central and eastern Africa. Unfortunately, the status of *Atractaspis microlepidota andersonii* Boulenger, 1905, known from the Near East and the Arabian Peninsula, was not discussed in that

Taxonomie

Als Grundlage unserer Klassifikation auf Gattungsebene und darunter dienen die Arbeit von David & Ineich (1999) sowie die verschiedenen Beiträge in Joger & Stümpel (2005) unter Berücksichtigung späterer Überarbeitungen sowie Stümpel & Joger (2009). Diese sind in den entsprechenden Abschnitten zu den Familien und Gattungen genauer spezifiziert. Die Klassifikation oberhalb der Gattungsebene fußt auf Vidal & Hedges (2002) sowie Vidal et al. (2007, 2008). Nach neuesten Untersuchungen gliedert sich die Familie Viperidae in die Unterfamilien Viperinae und Crotalinae.

ELAPIDAE Boie, 1827

Gattung *Bungarus* Daudin, 1803

In Erwartung der von Kuch angekündigten Revision erkennen wir hier für die abgegrenzte Region lediglich zwei Arten an. *Bungarus caeruleus* (Schneider, 1801) bedarf keiner weiteren Kommentare. Khan (1985) überarbeitete diese Art und definierte *Bungarus sindanus razai* Khan, 1985, deren Berechtigung jedoch fraglich bleibt (Vogel 2006). Diese Art wurde auch unlängst von Kuch (2004) diskutiert.

Gattung *Naja* Laurenti, 1768

Diese Gattung wurde von Wallach et al. (2009) überarbeitet und in vier Untergattungen aufgeteilt.

Lange als Unterart von *Naja naja* angesehen, wurde *Naja oxiana* (Eichwald, 1831) von Wüster & Thorpe (1991, 1992a-b) und Wüster (1996) schließlich Artrang zugebilligt (siehe auch Ananjeva et al. (2004, 2006). Die vormalige Unterart *Naja haje arabica* Scortecci, 1932, welche zu dem in erster Lienie afrikanischen *Naja haje*-Komplex gehört, wurde von Trape et al. (2009) in den Artrang erhoben.

Gattung *Walterinnesia* Lataste, 1887

Diese hauptsächlich im Nahen und Mittleren Osten heimische Gattung wurde von Nilson & Rastegar-Pouyani (2007) überarbeitet, wobei auch das Taxon *Walterinnesia morgani* (Mocquard, 1905) für die östlichen Populationen revalidiert wurde.

LAMPROPHIIDAE Fitzinger, 1843

Gattung *Atractaspis* Smith, 1849

Die verschlungene Taxonomie dieser weitestgehend afrikanischen Gattung wurde bereits von Dobiey & Vogel (2007) diskutiert. Lediglich zwei Arten sind aus der in diesem Buch behandelten Region bekannt, und bei beiden herrscht hinsichtlich ihrer systematischen Zuordnung nach wie vor Uneinigkeit. Wir folgen Vidal et al. (2007, 2008) und setzen *Atractaspis* in die Unterfamilie Atractaspidinae der Familie Lamprophiidae.

Atractaspis microlepidota Günther, 1866, wird seit langem als Vertreter der Fauna des Nahen Ostens geführt (eine diesbezügliche Übersicht findet sich bei David & Ineich 1999). Trape et al. (2006) unterzogen die Gruppe um *Atractaspis microlepidota* Günther, 1866, einer Überarbeitung, in deren Verlauf *A. microlepidota* als nur im Senegal, in Gambia und Mauretanien vorkommend beschränkt wurde. Diese Autoren revalidierten auch *Atractaspis watsoni* Boulenger, 1908, für die Populationen in Zentral- und Ostafrika. Leider wurde in diesem Zusammenhang nicht auch der Status von *Atractaspis microlepidota andersonii* Boulenger,

paper. According to the original description (BOULENGER, 1905), this taxon does not agree in the combination of lepidotic characters and colour pattern with any of the three species recognized by TRAPE et al. (2006). Pending a revision of eastern populations of the *A. microlepidota* complex, we here regard the taxon *andersonii* as a distinct species. We also consider *Atractaspis engaddensis* HAAS, 1950, described from Ein Gaddi, Israel, to be a valid species. It is distinct from both *A. microlepidota* and *A. andersonii* and occurs in sympatry with the latter species along the west coast of Arabia (see EGAN 2007).

VIPERIDAE OPPEL, 1811

We follow in part VIDAL et al. (2007), who recognized four subfamilies in the family Viperidae. Subsequently, WÜSTER et al. (2008) recognized only two subfamilies, Viperinae OPPEL, 1811 and Crotalinae OPPEL, 1811. The taxonomy of several genera is still in a state of flux, especially *Vipera, Echis* and *Gloydius*. Our generic allocations are explained in the respective generic and specific accounts. Major recent general overviews of vipers in the region covered in this volume can be found in JOGER & STÜMPEL (2005), DE SMEDT (2006) and especially STÜMPEL & JOGER (2009). Earlier works were discussed by DAVID & INEICH (1999).

Crotalinae OPPEL, 1811

Genus *Gloydius* HOGE & ROMANO HOGE, 1981

We here follow the specific and subspecific arrangements proposed by ORLOV & BARABANOV (2000a), with the exception of the subspecific nomenclature of *Gloydius halys* (PALLAS, 1776) (see below). GLOYD & CONANT (1990) stated *Gloydius blomhoffii* (BOIE, 1826) to occur on Sakhalin Island, but ORLOV & BARABANOV (2000a) pointed out that this taxon was absent from this and adjacent islands. The taxonomy of *Gloydius halys* (PALLAS, 1776) is still incompletely understood and under discussion. A summary of the subspecific classification was provided by DAVID & INEICH (1999). These authors accepted four subspecies. ORLOV & BARABANOV (2000a) and DIRKSEN & BÖHME (2005) recognized the same subspecific taxa, although not under the same names (see below) plus two more subspecies of uncertain status, one of which is furthermore not present in the region covered by the present book.

The major point of controversy revolves around the type locality of *Coluber halys* PALLAS, 1776. PALLAS (1776: 703) merely stated that this species occurred "*In aridissimis desert australis*" ("in the dryest southern deserts") without giving a more precise locality. Subsequently, STRAUCH (1873) restricted the type locality to Lugaskoi Factory, on the Ienissei River in Siberia. Seemingly, STRAUCH (1873) selected this locality on the basis of PALLAS (1814), in which PALLAS stated that "*Vipera halys* occurs in Upper Ienissei". In contrast, and based upon the accounts of other species in the same volume of PALLAS and the travel itinerary of this author, BOUR (1993) argued that this locality could not be in Siberia, as stated by previous authors, but should rather be pre-

1905, aus dem Nahen Osten und von der Arabischen Halbinsel beleuchtet. Der Erstbeschreibung (BOULENGER, 1905) zufolge stimmt dieses Taxon in seiner Kombination von Beschuppungs- und Färbungsmerkmalen mit keiner der drei von TRAPE et al. (2006) anerkannten Arten überein. In Erwartung einer Revision der östlichen Populationen des *A. microlepidota*-Komplexes sehen wir daher das Taxon *andersonii* als eigene Art an. Da die aus Ein Gaddi in Israel beschriebene *Atractaspis engaddensis* HAAS, 1950, sowohl von *A. microlepidota* als auch *A. andersonii* verschieden ist und zudem mit der zuletzt genannten Art an der Westküste von Arabien sympatrisch vorkommt (siehe EGAN 2007), betrachten wir sie hier ebenfalls als valide Art.

VIPERIDAE OPPEL, 1811

VIDAL et al. (2007) erkennen vier Unterfamilien der Viperidae an, welche nachfolgend von WÜSTER et al. (2008) auf zwei Unterfamilien reduziert wurden, Viperinae OPPEL, 1811 and Crotalinae OPPEL, 1811. Die Taxonomie mehrerer Gattungen ist noch immer nicht vollständig aufgeklärt, vor allem soweit es *Vipera, Echis* und *Gloydius* angeht. Unsere Gründe für die Zuordnung einzelner Arten zu bestimmten Gattungen sind in den nachfolgenden Abschnitten dargelegt. Maßgebliche allgemeine Übersichten über die in der in diesem Band behandelten Region heimischen Vipern finden sich bei JOGER & STÜMPEL (2005), DE SMEDT (2006) und vor allem STÜMPEL & JOGER (2009). Frühere Arbeiten wurden von DAVID & INEICH (1999) diskutiert.

Crotalinae OPPEL, 1811

Gattung *Gloydius* HOGE & ROMANO HOGE, 1981

Wir folgen hier der von ORLOV & BARABANOV (2000a) vorgeschlagenen Art- und Unterartgliederung mit Ausnahme der Unterartnomenklatur von *Gloydius halys* (PALLAS, 1776) (siehe unten). GLOYD & CONANT (1990) gaben ein Vorkommen von *Gloydius blomhoffii* (BOIE, 1826) auf der Insel Sachalin an, jedoch stellen ORLOV & BARABANOV (2000a) klar, dass dieses Taxon dort und auf den umliegenden Inseln fehlt.

Die Taxonomie von *Gloydius halys* (PALLAS, 1776) ist nach wie vor ungeklärt und wird noch diskutiert. Eine Zusammenfassung der Klassifikation der Unterarten wurde von DAVID & INEICH (1999) vorgelegt; diese Autoren erkannten vier Unterarten an. ORLOV & BARABANOV (2000a) und DIRKSEN & BÖHME (2005) akzeptierten ebenfalls diese vier Unterarttaxa, allerdings nicht unter denselben Namen (siehe unten) zuzüglich zwei weiterer, die jedoch nicht in der in diesem Buch behandelten Region vorkommen.

Anlass für die fortgesetzte Uneinigkeit gibt der Typusfundort von *Coluber halys* PALLAS, 1776. PALLAS (1776: 703) gab lediglich an, dass diese Art „*In aridissimis desert australis*" („in den sehr trockenen südlichen Wüsten") vorkäme, ohne einen genaueren Fundort zu benennen. In der Folge restringierte STRAUCH (1873) den Typusfundort auf Lugaskoi Fabrik am Fluss Jenissei in Sibirien. Anscheinend wählte STRAUCH (1873) diesen Ort auf der Grundlage von PALLAS (1814), in welcher PALLAS angab, dass „*Vipera halys* ... am oberen Jenissei" vorkomme. Im Gegensatz dazu und gemäß den Beschreibungen anderer Arten im selben Band von PALLAS und dem Reisetagebuch dieses Autors argumentierte BOUR (1993) hingegen, dass der Fundort nicht wie von vorangegangenen Auto-

sumed to lie in the Naryn Desert or Ryn Peski [Ryn Desert], towards today's border between Russia and Kazakhstan. Furthermore, the Naryn Desert was the only "very arid desert" through which PALLAS passed. ORLOV & BARABANOV (2000a), followed by ANANJEVA et al. (2004, 2006) and GUMPRECHT et al. (2004), rejected BOUR's (1993) interpretation and took STRAUCH's statement as a basis to designate a neotype from this area (Lugaskoi Zavod). It is very unfortunate that ORLOV & BARABANOV (2000a) then selected a specimen from another locality (84 km from Kyatha, Burin-Khan Mountain, Mongolia), at least 1,000 km east of Lugaskoi (see also ORLOV & BARABANOV 2000b). Whatever the reason(s) of PALLAS (1814) might have been to mention *Vipera halys* from the Upper Ienissei region, his publication issued in 1814 is definitely not the original description and hence the Siberian locality cannot be considered to be the type locality of *Coluber halys* PALLAS, 1776. We therefore regard as the valid type locality the one given in PALLAS (1776), which also has priority. As a consequence, BOUR's (1993) arguments have to be followed with regard to the type locality being located in the desert region of the Lower Volga River.

As was shown by BOUR (1993), the consequence is that the combination *Gloydius halys halys* (PALLAS, 1776) should be applied to western populations, with *Gloydius halys caraganus* (EICHWALD, 1831) as a synonym; the Siberian and Mongolian populations were named by BOUR (1993) *G. halys mogoi* (*Gloydius halys halys* sensu ORLOV & BARABANOV 2000a). This interpretation was accepted by DE SMEDT (2006). Thus, we here accept four subspecies for our covered area: *Gloydius halys halys* (PALLAS, 1776), *Gloydius halys boehmei* (NILSON, 1983), *Gloydius halys caucasicus* (NIKOLSKY, 1916), and *Gloydius halys mogoi* (BOUR, 1993). *Agkistrodon halys stejnegeri* (RENDAHL, 1933) was recognized by ORLOV & BARABANOV (2000a) as a valid subspecies for populations of northern China and extreme southeastern Mongolia. The status of this taxon is unsettled. It is probably a synonym of *Gloydius intermedius* (STRAUCH, 1868) and it is not recognized here.

The specific rank of *Gloydius intermedius* (STRAUCH, 1868) was confirmed by GLOYD & CONANT (1990), who recognized four subspecies. A discussion was then provided by DAVID & INEICH (1999), who considered this species to be monotypic. ORLOV & BARABANOV (2000a) presented sound arguments to regard *G. saxatilis* as a synonym of *G. intermedius*. This taxonomy was also adopted by ORLOV & BARABANOV (2000a), ANANJEVA et al. (2004, 2006), and GUMPRECHT et al. (2004).

Lastly, *Gloydius ussuriensis* (EMELIANOV, 1929), considered a subspecies of *Gloydius halys* or *G. intermedius* by numerous authors (see DAVID & INEICH 1999), was raised to full species status by TORIBA (1986). This position was rejected by SZCZERBAK (2003) but followed by DAVID & INEICH (1999), ORLOV & BARABANOV (2000a), ANANJEVA et al. (2004, 2006), and GUMPRECHT et al. (2004).

ren angegeben in Sibirien liegen könne, sondern eher in der Wüste Naryn oder Ryn Peski [Ryn-Wüste], nahe der heutigen Grenze zwischen Russland und Kasachstan zu suchen sei. Darüber hinaus ist die Wüste Naryn die einzige „sehr trockene Wüste", die von PALLAS durchquert wurde. ORLOV & BARABANOV (2000a) und nachfolgend auch ANANJEVA et al. (2004, 2006) und GUMPRECHT et al. (2004) wiesen BOURS (1993) Interpretation jedoch zurück und übernahmen STRAUCHS Angaben als Grundlage für die Festlegung eines Neotypus aus diesem Gebiet (Lugaskoi Zavod). Überaus unglücklich ist dabei, dass ORLOV & BARABANOV (2000a) ein Exemplar von einem anderen Fundort, nämlich von 84 km außerhalb von Kyatha, Berg Burin-Khan, Mongolei, auswählten, also mindestens 1000 km östlich von Lugaskoi gelegen (siehe auch ORLOV & BARABANOV 2000b). Welche Gründe auch immer PALLAS (1814) bewogen haben mögen, *Vipera halys* für die Region des oberen Jenissei anzugeben, seine Veröffentlichung von 1814 ist definitiv nicht die Erstbeschreibung, und aus diesem Grund kommt der Ort in Sibirien nicht als Typuslokalität von *Coluber halys* PALLAS, 1776, infrage. Wir sehen daher das bei PALLAS (1776) genannte Gebiet als den gültigen Typusfundort an, zumal dieser auch Priorität hat. Folglich muss auch der Argumentation von BOUR (1993) zur Lage des Typusfundortes in der Wüstenregion an der unteren Wolga Folge geleistet werden.

Wie von BOUR (1993) aufgezeigt, ergibt sich daraus die Notwendigkeit, die Namenskombination *Gloydius halys halys* (PALLAS, 1776) auf die westlichen Populationen anzuwenden, wodurch *Gloydius halys caraganus* (EICHWALD, 1831) zum Synonym wird. Die sibirischen und mongolischen Populationen wurden von BOUR (1993) als *Gloydius halys mogoi* neu benannt (*G. halys halys* sensu ORLOV & BARABANOV 2000a). Diese Interpretation wurde auch von DE SMEDT (2006) übernommen. Somit akzeptieren wir hier vier Unterarten: *Gloydius halys halys* (PALLAS, 1776), *Gloydius halys boehmei* (NILSON, 1983), *Gloydius halys caucasicus* (NIKOLSKY, 1916) und *Gloydius halys mogoi* (BOUR, 1993). *Agkistrodon halys stejnegeri* (RENDAHL, 1933) wurde von ORLOV & BARABANOV (2000a) als valide Unterart für die Populationen aus dem nördlichen China und dem südöstlichen Teil der Mongolei angesehen. Der Status dieses Taxons ist unklar. Es ist wahrscheinlich ein Synonym zu *Gloydius intermedius* (STRAUCH, 1868) und wird hier nicht anerkannt.

Der Artrang von *Gloydius intermedius* (STRAUCH, 1868) wurde von GLOYD & CONANT (1990) bestätigt, welche auch vier Unterarten anerkannten. DAVID & INEICH (1999) diskutierten diese Art und betrachteten sie als monotypisch. ORLOV & BARABANOV (2000a) präsentierten überzeugende Argumente dafür, dass *G. saxatilis* als Synonym von *G. intermedius* aufzufassen sei. Diese Taxonomie wurde dann auch von ORLOV & BARABANOV (2000a), ANANJEVA et al. (2004, 2006) und GUMPRECHT et al. (2004) übernommen.

Schlußendlich wurde *Gloydius ussuriensis* (EMELIANOV, 1929), die von zahlreichen Autoren wechselweise als Unterart von *Gloydius halys* oder *G. intermedius* aufgefasst worden war (siehe DAVID & INEICH 1999), von TORIBA (1986) in den Artrang erhoben. Diese Maßnahme wurde zwar von SZCZERBAK (2003) abgelehnt, von DAVID & INEICH (1999), ORLOV & BARABANOV (2000a), ANANJEVA et al. (2004, 2006) und GUMPRECHT et al. (2004) jedoch akzeptiert.

Viperinae Oppel, 1811
Genus *Bitis* Gray, 1842
This largely African genus barely enters the area covered here in the form of the species *Bitis arietans* (Merrem, 1820). It should be noted that Arabian populations are referred to the nominal subspecies and not to *Bitis arietans somalica* Parker, 1949, which is limited to Somalia and northern Kenya. Nevertheless there are differences in colour and size between the Arabian and African populations.

Genus *Cerastes* Laurenti, 1768
Following the revisions by Werner et al. (1991, 1999) and Werner & Sivan (1992), this genus now includes three species (see also Stümpel & Joger, 2009). Werner et al. (1991) raised to full species status the former subspecies *Cerastes cerastes gasperettii* Leviton & Anderson, 1967. Werner et al. (1999) subsequently described a new subspecies from southwest Arabia, *Cerastes cerastes hoofieni* Werner, Sivan, Kushnir & Motro, 1999. This arrangement was adopted by Mallow et al. (2003). Werner et al. (1999) furthermore described a new subspecies of *Cerastes gasperettii*, *Cerastes gasperettii mendelssohni* Werner, Sivan, Kushnir & Motro, 1999, endemic to the 'Arava Valley of Israel. These taxa were accepted by Mallow et al. (2003).

Genus *Daboia* Gray, 1842
Obst (1983) resurrected *Daboia* for the species *Vipera russelii* (Shaw & Nodder, 1797), *V. palaestinae* Werner, 1838, and species of the *Vipera lebetina* and *V. xanthina* groups. Herrmann et al. (1992) restricted the genus *Daboia* to *Vipera russelii* and resurrected the genus *Macrovipera* Reuss, 1927 (see below) to accommodate species of the group of *Vipera lebetina* (see David & Ineich 1999). In contrast, Lenk et al. (2001) suggested assigning to *Daboia* the species then known as *Macrovipera palaestinae* as well as the African species *Macrovipera mauritanica* and *M. deserti* (see Dobiey & Vogel 2007). Subsequently, if the genus *Daboia* was widely recognized for *D. russelii* and *D. palaestinae*, Joger & Stümpel (2005) confirmed the validity of the genus *Macrovipera* for all members of the *lebetina*-group. This position was followed by subsequent authors (see Stümpel & Joger 2009).

Genus *Echis* Merrem, 1820
A first step towards a global revision of the difficult systematics of this genus appeared in Cherlin (1990) and Cherlin & Borkin (1990). However, many of taxa that these authors described were controversial and were not recognized by subsequent authors ; see the discussion in Herrmann & Joger (1996) and David & Ineich (1999). Recently, Pooke et al. (2009) and Arnold et al. (2009) addressed the taxonomy and phylogeny of the genus *Echis*. We here follow largely the taxonomic scheme at specific and subspecific levels proposed by these authors, in agreement with Stümpel & Joger (2009).
Cherlin (1990) divided *Echis pyramidum* (Geoffroy Saint-Hilaire, 1827) into three species (see David & Ineich, 1999), but this arrangement has not met with general acceptance. Pooke et al. (2009) showed that populations of "*Echis pyramidum*" from the Arabian Peninsula were genetically

Viperinae Oppel, 1811
Gattung *Bitis* Gray, 1842
Diese weitestgehend afrikanische Gattung erreicht das hier behandelte Gebiet nur knapp in Gestalt der Art *Bitis arietans* (Merrem, 1820). Betont werden muss, dass die arabischen Populationen der Nominatform zugeschlagen werden müssen und nicht *Bitis arietans somalica* Parker, 1949, welche auf Somalia und Nord-Kenia beschränkt ist.

Gattung *Cerastes* Laurenti, 1768
Nach den Revisionen seitens Werner et al. (1991, 1999) und Werner & Sivan (1992) enthält diese Gattung nunmehr drei Arten (siehe auch Stümpel & Joger, 2009). Werner et al. (1991) erhoben die frühere Unterart *Cerastes cerastes gasperettii* Leviton & Anderson, 1967, in den Artrang. Werner et al. (1999) beschrieben dann eine neue Unterart aus dem Südwesten Arabiens, *Cerastes cerastes hoofieni* Werner, Sivan, Kushnir & Motro, 1999. Diese Einteilung wurde von Mallow et al. (2003) übernommen. Werner et al. (1999) definierten weiterhin eine neue Unterart von *Cerastes gasperettii*, nämlich *Cerastes gasperettii mendelssohni* Werner, Sivan, Kushnir & Motro, 1999, die im Arava Tal in Israel endemisch ist. Diese Taxa wurden ebenfalls von Mallow et al. (2003) anerkannt.

Gattung *Daboia* Gray, 1842
Obst (1983) revalidierte *Daboia* für die Arten *Vipera russelii* (Shaw & Nodder, 1797), *V. palaestinae* Werner, 1838, und die Arten der *Vipera lebetina*- und *V. xanthina*-Gruppen.
Herrmann et al. (1992) beschränkten die Gattung *Daboia* auf die Art *Vipera russelii* und revalidierten die Gattung *Macrovipera* Reuss, 1927 (siehe unten) für die Arten der *Vipera lebetina*-Gruppe (siehe David & Ineich 1999). Im Gegensatz dazu schlugen Lenk et al. (2001) vor, die bisher als *Macrovipera palaestinae* sowie die afrikanischen Arten *Macrovipera mauritanica* und *M. deserti* *Daboia* zuzuschlagen (siehe Dobiey & Vogel 2007). Schließlich, nachdem die Gattung *Daboia* im Allgemeinen für *D. russelii* und *D. palaestinae* akzeptiert wurde, bestätigten Joger & Stümpel (2005) die Gültigkeit der Gattung *Macrovipera* für alle Formen der *lebetina*-Gruppe. Dieser Einteilung folgten andere Autoren (siehe Stümpel & Joger 2009).

Gattung *Echis* Merrem, 1820
Erste Schritte für umfassende Revisionen dieser Gattung gehen auf Cherlin (1990) und Cherlin & Borkin (1990) zurück. Allerdings erwiesen sich etliche der von diesen Autoren beschriebenen Taxa als strittig und wurden von nachfolgenden Autoren nicht anerkannt. Siehe auch die Diskussionen von Herrmann & Joger (1996) und David & Ineich (1999). Vor Kurzem handelten schließlich Pooke et al. (2009) sowie Arnold et al. (2009) die Taxonomie und Phylogenie der Gattung *Echis* ab. In Übereinstimmung mit Stümpel & Joger (2009) folgen wir hier dem taxonomischen Schema dieser Autoren.
Cherlin (1990) teilte *Echis pyramidum* (Geoffroy Saint-Hilaire, 1827) in drei Arten auf (see David & Ineich, 1999). Diese Aufteilung wurde jedoch überwiegend abgelehnt. Pooke et al. (2009) zeigten, dass die Populationen von "*Echis pyramidum*" der arabischen Halbinsel genetisch unterschiedlich von den afrikanischen Populationen sind. Diese Autoren re-

distinct from African populations. As a consequence, these latter authors revalidated *Echis borkini* CHERLIN, 1990 for populations of western Yemen and southwestern Arabia, and confirmed the validity of *Echis khosatzkii* CHERLIN, 1990 for populations of eastern Yemen and Sultanate of Oman (see below). This latter species has largely been considered a synonym of *Echis pyramidum* by most authors after CHERLIN (1190) but SCHÄTTI (2001) had previously considered this taxon to be valid, a view now supported by POOKE et al. (2009). *Echis pyramidum* is now restricted to Africa.

Regarding *Echis carinatus* (SCHNEIDER, 1801), we here follow AUFFENBERG & REHMAN (1991), GOLAY et al. (1993), and SCHÄTTI & GASPERETTI (1994) for the subspecific classification of this species, modified by POOKE et al. (2009). We therefore regard *Echis sochureki* STEMMLER, 1969 as a subspecies of *Echis carinatus*. Following POOKE et al. (2009), *Echis multisquamatus* CHERLIN, 1981, considered either a full species by ANANJEVA et al. (2004, 2006) or a subspecies of *E. carinatus* by most other authors, becomes a synonym of *E. carinatus sochureki*.

Echis coloratus (GÜNTHER, 1878) has been revised by BABOCSAY (2003), who also described a new subspecies, *Echis coloratus terraesanctae*, and a new related species, *Echis omanensis*. This latter taxon was recognized by POOKE et al. (2009) but these authors did not accept the validity of *E. c. terraesanctae*. However, they only briefly discussed the group of *Echis coloratus*. Pending an additional revision of this group, we retain *Echis coloratus terraesanctae* as a valid subspecies.

Genus *Macrovipera* REUSS, 1927

This genus was resurrected from the synonymy of the genus *Vipera* by HERRMANN et al. (1992) to include large Eurasian and North African vipers.

LENK et al. (2001) referred *Macrovipera palaestinae* and North African taxa of the complex of *Macrovipera lebetina* (LINNAEUS, 1758) to the genus *Daboia*, leaving the genus *Macrovipera* only with the two Asian species of the *lebetina* group. MALLOW et al. (2003) retained the genus *Macrovipera* for all members of the *lebetina* complex. JOGER & STÜMPEL (2005) and STÜMPEL & JOGER (2009) confirmed this interpretation.

The subspecific classification of *Macrovipera lebetina* is still under discussion. *Vipera lebetina schweizeri* (WERNER, 1935) was afforded specific rank by NILSON & ANDRÉN (1988a), a notion accepted by subsequent authors but not by STÜMPEL & JOGER (2009) who retained a subspecific status for the taxon *schweizeri*.

Furthermore, NILSON & ANDRÉN (1988b) described *Vipera lebetina transmediterranea* from North Africa and recognized *Vipera lebetina lebetina*, *M. lebetina obtusa*, and *M. lebetina turanica*. This arrangement was adopted by HERRMANN et al. (1992), who placed this group into the genus *Macrovipera*, and by subsequent authors (see GOLAY et al. 1993, DAVID & INEICH 1999). *Macrovipera lebetina euphratica* (MARTIN, 1838) was considered to be a synonym of *M. lebetina obtusa* by JOGER (1984) and most subsequent authors. CHIKIN & SZCZERBAK (1992) then recognized the validity of the subspecies *euphratica* and described *Macrovipera lebetina cernovi* for populations of southwestern Central Asia (Turkme-

validierten den Namen *Echis borkini* CHERLIN, 1990 für Populationen des westlichen Jemen und Südwest-Arabien und bestätigten die Gültigkeit des Namens *Echis khosatzkii* CHERLIN, 1990 für Populationen des östlichen Jemen und des Oman (siehe weiter unten). Der Name *Echis pyramidum* beschränkt sich nun auf afrikanische Populationen. *Echis khosatzkii* CHERLIN, 1990, welche auch als Folge der Aufspaltung von *Echis pyramidum* beschrieben wurde, wurde dann mit *E. pyramidum pyramidum* synonymisiert. SCHÄTTI (2001) hielt dagegen dieser Art für gültig, was von POOKE et al. (2009) bestätigt wurde.

Bezüglich *Echis carinatus* (SCHNEIDER, 1801) folgen wir hier AUFFENBERG & REHMAN (1991), GOLAY et al. (1993) und SCHÄTTI & GASPERETTI (1994) in der Aufspaltung der Unterarten, modifiziert nach POOKE et al. (2009). Daher betrachten wir *Echis sochureki* STEMMLER, 1969 als eine Unterart von *Echis carinatus*. Nach POOKE et al. (2009) wird *Echis multisquamatus* CHERLIN, 1981, welche zum Beispiel von ANANJEVA et al. (2004, 2006) als eigenständige Art oder von den meisten anderen Autoren als Unterart von *E. carinatus* betrachtet wurde, zum Synonym von *E. carinatus sochureki*.

Echis coloratus (GÜNTHER, 1878) wurde von BABOCSAY (2003) revidiert, wobei auch die neue Unterart *Echis coloratus terraesanctae*, und eine neue verwandte Art, *Echis omanensis*. beschrieben wurden. *Echis omanensis* wurde von POOKE et al. (2009) im Gegensatz zu *E. c. terraesanctae* akzeptiert. Da die *Echis coloratus*-Gruppe von diesen Autoren nur sehr kurz diskutiert wird, betrachten wir bis zu einer abschließenden Bearbeitung *Echis coloratus terraesanctae* als eine valide Unterart.

Gattung *Macrovipera* REUSS, 1927

Diese Gattung wurde von HERRMANN et al. (1992) aus der Synonymie der Gattung *Vipera* herausgenommen, um die großen Vipern Eurasiens und Nordafrikas aufzunehmen.

LENK et al. (2001) schlugen die nordafrikanischen Taxa und *Macrovipera palaestinae* der Gattung *Daboia* zu, wodurch die Gattung *Macrovipera* auf lediglich zwei Arten aus der *Macrovipera lebetina*-Gruppe reduziert wurde. MALLOW et al. (2003) verwendeten die Gattung *Macrovipera* für alle Formen des *lebetina*-Komplexes. JOGER & STÜMPEL (2005) und STÜMPEL & JOGER (2009) bestätigten diese interpretation.

Die Unterart-Klassifikation von *Macrovipera lebetina* wird noch diskutiert. *Vipera lebetina schweizeri* (WERNER, 1935) wurde von NILSON & ANDRÉN (1988a) Artrang zugesprochen, eine Einschätzung, die von vielen nachfolgenden Autoren übernommen wurde, nicht aber von STÜMPEL & JOGER (2009).

Weiterhin beschrieben NILSON & ANDRÉN (1988b) *Vipera lebetina transmediterranea* aus Nordafrika und erkannten *Vipera lebetina lebetina*, *M. lebetina obtusa* und *M. lebetina turanica* an. Diese Einteilung wurde von HERRMANN et al. (1992) übernommen (die diese Gruppe in die Gattung *Macrovipera* überstellten) ebenso wie auch von späteren Autoren (siehe GOLAY et al. 1993, DAVID & INEICH 1999). *Macrovipera lebetina euphratica* (MARTIN, 1838) wurde von JOGER (1984) und den meisten nachfolgenden Autoren als Synonym von *M. lebetina obtusa* angesehen. CHIKIN & SZCZERBAK (1992) erkannten dann die Gültigkeit der Unterart *euphratica* und beschrieben *Macrovipera lebetina cernovi* für die Populationen im Südwesten Zentralasiens (Turkmenien und Usbekistan), Iran, Afghanistan

nia and Uzbekistan), Iran, Afghanistan and Kashmir (Pakistan or India). This interpretation was maintained by DAVID et al. (1999), MALLOW et al. (2003), ANANJEVA et al. (2004, 2006), SZCZERBAK & BÖHME (2005) and DE SMEDT (2006). However, the form occurring in southern Turkey, Lebanon and Iraq is distinct from *M. lebetina obtusa* (A. TEYNIÉ, pers. comm., April 2008), and the name *euphratica* is available for it, at least for a portion of this population if not all of it. Accordingly, we here accept the validity of *Macrovipera lebetina euphratica* (MARTIN, 1838). This position was also adopted by STÜMPEL & JOGER (2009) although it is still tentative. Furthermore, according to NILSON (cit. fide DAVID & INEICH 1999) and SZCZERBAK & BÖHME (2005), populations from Afghanistan, Pakistan and India, currently referred to the subspecies *cernovi*, may represent a distinct taxon for which the name *Vipera peilei* MURRAY, 1892, is available. Pending a revision of these populations on the eastern limits of the range, we retain them under the subspecific name *M. l. cernovi*. Thus, *Macrovipera lebetina* currently comprises six subspecies, five of which occur within the region covered by the present volume (see also STÜMPEL & JOGER, 2009): *Macrovipera lebetina lebetina*, *M. lebetina cernovi* CHIKIN & SZCZERBAK, 1992, *M. lebetina euphratica* (MARTIN, 1838), *M. lebetina obtusa* (DWIGUBSKY, 1832), and *M. lebetina turanica* (CHERNOV *in* TERENT'EV & CHERNOV, 1940).

Genus *Montivipera* NILSON, TUNIYEV, ANDRÉN, ORLOV, JOGER & HERRMANN, 1999

NILSON et al. (1999) originally defined *Montivipera* as a subgenus of *Vipera* to accommodate montane species of the *V. xanthina-raddei* group of Turkey and the Near East. LENK et al. (2001) and GARRIGUES et al. (2005) later presented data warranting the recognition of this subgenus as a full genus, a view that was adopted by JOGER (2005) and STÜMPEL & JOGER (2009), and which is also followed here. In contrast, MALLOW et al. (2003) and DE SMEDT (2006) maintained its original subgeneric status. Currently, this Eurasian genus includes the eight species treated here. These species of mountain vipers were reviewed by NILSON & ANDRÉN (1986, 1992, 1996). However, not all of them are unanimously accepted.

Montivipera albicornuta (NILSON & ANDRÉN, 1985) and *Montivipera latifii* (MERTENS, DAREVSKY & KLEMMER, 1967) were considered synonyms of *Vipera raddei* BOETTGER, 1890 by SCHÄTTI et al. (1991, 1992), GOLAY et al. (1993), and SCHWEIGER (1995). We here follow NILSON & ANDRÉN (1992), NILSON et al. (1999), LENK et al. (2001), and STÜMPEL & JOGER (2009) who regarded these taxa as valid species. This arrangement was followed by MALLOW et al. (2003).

On the other hand, SCHÄTTI et al. (1991, 1992) considered *Montivipera albizona* (NILSON, ANDRÉN & FLÄRDH, 1990), *Montivipera bornmuelleri* (WERNER, 1898) and *Montivipera bulgardaghica* (NILSON & ANDRÉN, 1985) to be synonyms of *Vipera xanthina* (GRAY, 1849). We follow NILSON & ANDRÉN (1992), NILSON et al. (1999), LENK et al. (2001) and STÜMPEL & JOGER (2009), in accepting all these biologically isolated montane taxa as valid species, as did MALLOW et al. (2003).

und Kaschmir (Pakistan bzw. Indien). Diese Einteilung wurde von DAVID et al. (1999), MALLOW et al. (2003), ANANJEVA et al. (2004, 2006), SZCZERBAK & BÖHME (2005) und DE SMEDT (2006) übernommen. Allerdings unterscheidet sich die im Süden der Türkei, im Libanon und Irak vorkommende Form von *M. lebetina obtusa* (A. TEYNIÉ, pers. Mitt., April 2008) von dieser, und der Name *euphratica* steht für diese Form zur Verfügung, zumindest für einen Teil dieser Population wenn nicht für ihre Gesamtheit. Folglich akzeptieren wir hier die Gültigkeit von *Macrovipera lebetina euphratica* (MARTIN, 1838). Dieser Auffassung sind auch STÜMPEL & JOGER (2009) auch wenn diese Unterteilung nicht endgültig sein mag.

Weiterhin sind NILSON (cit. fide DAVID & INEICH 1999) und SZCZERBAK & BÖHME (2005) zufolge die Populationen in Afghanistan, Pakistan und Indien, welche zur Zeit der Unterart *cernovi* zugerechnet werden, möglicherweise ein separates Taxon, für das der Name *Vipera peilei* MURRAY, 1892, zur Verfügung steht. In Erwartung einer Revision dieser Populationen am Ostrand der Verbreitung behalten wir auf Unterartebene den Namen *M. l. cernovi* bei. Somit umfasst *Macrovipera lebetina* zum gegenwärtigen Zeitpunkt sechs Unterarten, von denen fünf in der vom vorliegenden Band betroffenen Region vorkommen: *Macrovipera lebetina lebetina*, *M. lebetina cernovi* CHIKIN & SZCZERBAK, 1992, *Macrovipera lebetina euphratica* (MARTIN, 1838), *M. lebetina obtusa* (DWIGUBSKY, 1832) und *M. lebetina turanica* (CHERNOV *in* TERENT'EV & CHERNOV, 1940).

Gattung *Montivipera* NILSON, TUNIYEV, ANDRÉN, ORLOV, JOGER & HERRMANN, 1999

NILSON et al. (1999) definierten *Montivipera* ursprünglich als Untergattung der Gattung *Vipera*, in welcher die montanen Arten der *V. xanthina-raddei*-Gruppe aus der Türkei und dem Nahen Osten zusammengefasst wurden. LENK et al. (2001) und GARRIGUES et al. (2005) legten später Daten vor, die eine Aufwertung der Untergattung zur Gattung rechtfertigten. Diese Ansicht wurde von JOGER (2005) und STÜMPEL & JOGER (2009) übernommen und wird auch hier verwendet, wurde aber von MALLOW et al. (2003) und DE SMEDT (2006) abgelehnt. Gegenwärtig umfasst diese eurasische Gattung die hier behandelten acht Arten. Diese Gruppe von Bergottern wurde von NILSON & ANDRÉN (1986, 1992, 1996) bearbeitet. Allerdings werden nicht alle einhellig akzeptiert (siehe die entsprechenden folgenden Abschnitte).

Montivipera albicornuta (NILSON & ANDRÉN, 1985) und *Montivipera latifii* (MERTENS, DAREVSKY & KLEMMER, 1967) wurden von SCHÄTTI et al. (1991, 1992), GOLAY et al. (1993) und SCHWEIGER (1995) als Synonyme von *Vipera raddei* BOETTGER, 1890, angesehen. Wir folgen hier jedoch NILSON & ANDRÉN (1992), NILSON et al. (1999), LENK et al. (2001), MALLOW et al. (2003) sowie STÜMPEL & JOGER (2009), die diese Taxa als valide ansehen.

Andererseits betrachteten SCHÄTTI et al. (1991, 1992) *Montivipera albizona* (NILSON, ANDRÉN & FLÄRDH, 1990), *Montivipera bornmuelleri* (WERNER, 1898) und *Montivipera bulgardaghica* (NILSON & ANDRÉN, 1985) als Synonyme von *Vipera xanthina* (GRAY, 1849). Wir folgen hier wiederum NILSON & ANDRÉN (1992), NILSON et al. (1999), LENK et al. (2001), MALLOW et al. (2003) sowie STÜMPEL & JOGER (2009) und akzeptieren alle diese biologisch isolierten Gebirgsformen als valide Arten.

Lastly, Schätti et al. (1991, 1992) suggested that *Montivipera wagneri* (Nilson & Andrén, 1984) was conspecific with *Vipera xanthina*. Like Mallow et al. (2003), we follow Joger et al. (1988), Nilson & Andrén (1992), Nilson et al. (1999), and Lenk et al. (2001) who all regarded this biologically isolated montane form as a valid species.

Schließlich schlugen Schätti et al. (1991, 1992) vor, dass *Montivipera wagneri* (Nilson & Andrén, 1984) mit *Vipera xanthina* konspezifisch sei. Wie Mallow et al. (2003) folgen wir hier jedoch Joger et al. (1988), Nilson & Andrén (1992), Nilson et al. (1999) und Lenk et al. (2001), die in dieser biologisch isolierten Montanform eine valide Art sahen.

Genus *Pseudocerastes* Boulenger, 1896

This genus was reviewed by Bostanchi et al. (2006), whom we follow here. Bostanchi et al. (2006) separated as a distinct species *Pseudocerastes fieldii* Schmidt, 1930, a taxon traditionally viewed as a subspecies of *Pseudocerastes persicus* (Duméril, Bibron & Duméril, 1854). This now leaves the latter taxon as monotypic. Bostanchi et al. (2006) also described *Pseudocerastes urarachnoides* Bostanchi, Anderson, Kami & Papenfuss, 2006, a rare species with a very unusual tail tip. Lastly, we regard the population from the lava fields in Syria as a potentially distinct taxon and list it as *Pseudocerastes* cf. *fieldii*, pending further investigations.

Gattung *Pseudocerastes* Boulenger, 1896

Diese Gattung wurde von Bostanchi et al. (2006) überarbeitet, deren Ergebnissen wir uns hier anschließen. Traditionell als eine Unterart von *Pseudocerastes persicus* (Duméril, Bibron & Duméril, 1854) aufgefasst, betrachten Bostanchi et al. (2006) das Taxon *Pseudocerastes fieldii* Schmidt, 1930, als eigenständige Art. Damit wird das erstgenannte Taxon monotypisch. Bostanchi et al. (2006) beschrieben weiterhin mit *Pseudocerastes urarachnoides* Bostanchi, Anderson, Kami & Papenfuss, 2006, eine Art mit einer höchst ungewöhnlichen Schwanzspitze. Wir betrachten die Population der Lavafelder in Syrien als eine potenziell unterschiedliche Form und bezeichnen sie vorläufig als *Pseudocerastes* cf. *fieldii*.

Genus *Vipera* Laurenti, 1768

At generic level, various classification models have been proposed over the last decades. At specific level, controversy continues to exist with regard to several species; these are detailed below in the respective species accounts. An excellent overview of the contents and biology of the genus *Vipera* was provided by De Smedt (2006).

The subspecific content of *Vipera ammodytes* (Linnaeus, 1758) has not yet been resolved. Up to six subspecies have been recognized by some authors. We follow Biella & Blätter (1992) and Golay et al. (1993) who considered *Vipera ammodytes ruffoi* Bruno, 1968 (a subspecies accepted by De Smedt 2006) and *V. ammodytes gregorwallneri* Sochurek, 1974, to be synonyms of *V. ammodytes ammodytes*. This notion was supported by Ursenbacher et al. (2008) on the basis of molecular analyses. In agreement with Heckes et al. (2005), we regard *Vipera ammodytes montandoni* Boulenger, 1904, as a valid subspecies. On the other hand, Ursenbacher et al. (in press) confirm the subspecific status of *Vipera ammodytes transcaucasiana*, a taxon regarded as a valid species by Obst (1983), Herrmann et al. (1992), Nilson et al. (1999), Mallow et al. (2003), Szczerbak (2003), Ananjeva et al. (2004, 2006), and De Smedt (2006). As a consequence, and following Heckes et al. (2005), we accept here only four subspecies: *Vipera ammodytes ammodytes* (Linnaeus, 1758), *Vipera ammodytes meridionalis* Boulenger, 1903, *Vipera ammodytes montandoni* Boulenger, 1904, and *Vipera ammodytes transcaucasiana* Boulenger, 1913.

Vipera anatolica Eiselt & Baran, 1970, a member of the *Vipera ursinii* complex, was recognized at specific level by Nilson & Andrén (2001) and Mallow et al. (2003).

The subspecific arrangement of *Vipera aspis* (Linnaeus, 1758) is still highly controversial. Saint Girons (1978), Saint Girons et al. (1983), and Golay et al. (1993) recognized five subspecies. Subsequently, Zuffi (2002) proposed

Gattung *Vipera* Laurenti, 1768

Auf Gattungsebene sind über die letzten Jahrzehnte verschiedene Klassifikationsmodelle vorgeschlagen worden. Hinsichtlich der Arten besteht in einigen Fällen nach wie vor Uneinigkeit; diese werden in den entsprechenden nachfolgenden Abschnitten erläutert. Eine sehr gute Übersicht über den Umfang und die Biologie der Gattung *Vipera* wurde von De Smedt (2006) vorgelegt.

Die Gliederung der Unterarten von *Vipera ammodytes* (Linnaeus, 1758) ist bislang nicht umfassend gelöst worden. Von verschiedenen Autoren werden bis zu sechs Unterarten anerkannt. Wir folgen hier jedoch Biella & Blätter (1992) und Golay et al. (1993), welche *Vipera ammodytes ruffoi* Bruno, 1968 (eine von De Smedt 2006 akzeptierte Unterart) und *V. ammodytes gregorwallneri* Sochurek, 1974, als Synonyme von *V. ammodytes ammodytes* auffassen. Diese Einstellung wird auch von Ursenbacher et al. (2008) auf der Grundlage von molekulargenetischen Untersuchungen gestützt. In Übereinstimmung mit Heckes et al. (2005) erkennen wir *Vipera ammodytes montandoni* Boulenger, 1904, als valide Unterart an. Andererseits bestätigen Ursenbacher et al. (2008) den Unterartstatus von *Vipera ammodytes transcaucasiana*, einem Taxon, das von Obst (1983), Herrmann et al. (1992), Nilson et al. (1999), Mallow et al. (2003), Szczerbak (2003), Ananjeva et al. (2004, 2006) und De Smedt (2006) als valide Art angesehen wurde. Als Konsequenz daraus und in Übereinstimmung mit Heckes et al. (2005) akzeptieren wir hier lediglich vier Unterarten: *Vipera ammodytes ammodytes* (Linnaeus, 1758), *Vipera ammodytes meridionalis* Boulenger, 1903, *Vipera ammodytes montandoni* Boulenger, 1904 und *Vipera ammodytes transcaucasiana* Boulenger, 1913.

Vipera anatolica Eiselt & Baran, 1970, ein Vertreter des *Vipera ursinii*-Komplexes, wurde von Nilson & Andrén (2001) und Mallow et al. (2003) als Art anerkannt.

Die Gliederung von *Vipera aspis* (Linnaeus, 1758) in Unterarten ist noch immer überaus strittig. Saint Girons (1978), Saint Girons et al. (1983) und Golay et al. (1993) übernahmen fünf Unterarten. In der Folge schlug Zuffi (2002) eine erheblich verän-

a substantially modified taxonomy of this taxon, in elevating to full species status *Vipera aspis zinnikeri* KRAMER, 1958 and *Vipera hugyi* SCHINZ, 1834. This arrangement was largely refuted by GARRIGUES et al. (2005). We follow here GOLAY (2005) and URSENBACHER et al. (2006) in referring *Vipera aspis atra* MEISNER, 1820, to the synonymy of *Vipera aspis aspis* (LINNAEUS, 1758). However, we will refer to this taxon as "form *atra*", due to its largely melanistic coloration. Lastly, TRUTNAU et al. (2005) regarded six subspecies as valid, namely *V. aspis aspis* (LINNAEUS, 1758), *V. aspis atra* MEISNER, 1820, *V. a. francisciredi* LAURENTI, 1768, *V. aspis hugyi* SCHINZ, 1834, *Vipera aspis montecristi* MERTENS, 1956 (here considered a synonym of *V. a. hugyi*), and *Vipera aspis zinnikeri* KRAMER, 1958. MALLOW et al. (2003) and DE SMEDT (2006) did not regard *V. aspis montecristi* as valid and hence accepted only five subspecies. Pending a revision of the entire *V. aspis* complex, we maintain a conservative arrangement and regard as valid only four subspecies: *Vipera aspis aspis*, *Vipera aspis francisciredi*, *Vipera aspis hugyi*, and *Vipera aspis zinnikeri*.

The subspecific classification of *Vipera berus* (LINNAEUS, 1758) is still controversial as well. The subspecies *Vipera berus sachalinensis* TSAREWSKY, 1916, was granted specific status by JOGER et al. (1995) and treated as such by JOGER et al. (1997), ADNAGULOV et al. (2000) and ANANJEVA et al. (2004, 2006). According to JOGER *in* DAVID & INEICH (1999) this taxonomic change was unjustified, a view subsequently shared by KALYABINA-HAUF et al. (2004). Furthermore, JOGER pointed out that *Vipera nikolskii* VEDMEDERJA, GRUBANT & RUDAYEVA, 1986, should be regarded as a subspecies of *V. berus*. This notion was discussed but not fully resolved by BAKIEV et al. (2005); it was not accepted by either MALLOW et al. (2003) or DE SMEDT (2006) who regarded *Vipera nikolskii* as a valid species. The *Vipera berus* complex was revised by KALYABINA-HAUF et al. (2004) and extensively reviewed by NILSON et al. (2005). Their arrangements are followed here, with the recognition of four subspecies: *Vipera berus berus* (LINNAEUS, 1758), *Vipera berus bosniensis* BOETTGER *in* MOJSISOVICS, 1889, *Vipera berus nikolskii* VEDMEDERJA, GRUBANT & RUDAYEVA, 1986, and *Vipera berus sachalinensis* TSAREWSKY, 1916. JOGER *in* DAVID & INEICH (1999) suggested that the taxon *bosniensis* might be deserving of specific status.

One of the most poorly known vipers is *Vipera darevskii* VEDMEDERJA, ORLOV & TUNIYEV, 1986. The morphology, distribution, relationships and status of this rare species have been discussed by GENIEZ & TEYNIÉ (2005).

Vipera dinniki NIKOLSKY, 1913, a taxon long confused with *Vipera kaznakovi* NIKOLSKY, 1909, was discussed by ORLOV & TUNIYEV (1986, 1990), NILSON et al. (1994, 1995) and ORLOV & TUNIYEV (2005).

Vipera ebneri KNOEPFFLER & SOCHUREK, 1955, was considered a synonym of *Vipera ursinii eriwanensis* (REUSS, 1933) (see DAVID & INEICH 1999). NILSON & ANDRÉN (2001) resurrected it at species level, a concept subsequently accepted by MALLOW et al. (2003), but JOGER & DELY (2005) continued to view *Vipera ebneri* as a subspecies of *V. renardi*.

derte systematische Einteilung für dieses Taxon vor, in der *Vipera aspis zinnikeri* KRAMER, 1958 und *Vipera hugyi* SCHINZ, 1834, Artstatus zugebilligt wurde. Dieses Arrangement wurde von GARRIGUES et al. (2005) weitgehend abgelehnt. Wir folgen hier GOLAY (2005) und URSENBACHER et al. (2006) und stellen *Vipera aspis atra* MEISNER, 1820, in die Synonymie von *Vipera aspis aspis* (LINNAEUS, 1758). Trotzdem verweisen wir auf dieses Taxon wegen der weitestgehend melanistischen Färbung als „form *atra*". Schließlich sahen TRUTNAU et al. (2005) sechs Unterarten als gültig an, nämlich *V. aspis aspis* (LINNAEUS, 1758), *V. aspis atra* MEISNER, 1820, *V. a. francisciredi* LAURENTI, 1768, *V. aspis hugyi* SCHINZ, 1834, *Vipera aspis montecristi* MERTENS, 1956 (hier als Synonym von *V. a. hugyi* betrachtet) und *Vipera aspis zinnikeri* KRAMER, 1958. MALLOW et al. (2003) und DE SMEDT (2006) betrachteten *V. aspis montecristi* nicht als valide und akzeptierten somit nur fünf Unterarten. In Erwartung einer Revision des gesamten *V. aspis*-Komplexes behalten wir eine konservative Einteilung bei und behandeln nur vier Unterarten als gültig: *Vipera aspis aspis*, *Vipera aspis francisciredi*, *Vipera aspis hugyi* und *Vipera aspis zinnikeri*.

Die Klassifikation der Unterarten von *Vipera berus* (LINNAEUS, 1758) ist ebenfalls nach wie vor kontrovers. Die Unterart *Vipera berus sachalinensis* TSAREWSKY, 1916, wurde von JOGER et al. (1995) in den Artrang erhoben und von JOGER et al. (1997), ADNAGULOV et al. (2000) und ANANJEVA et al. (2004, 2006) als solche behandelt. Gemäß JOGER in DAVID & INEICH (1999) war diese taxonomische Veränderung jedoch ungerechtfertigt, eine Ansicht, die in der Folge auch von KALYABINA-HAUF et al. (2004) geteilt wurde. Weiterhin machte JOGER deutlich, dass *Vipera nikolskii* VEDMEDERJA, GRUBANT & RUDAYEVA, 1986, als Unterart von *V. berus* angesehen werden sollte. Diese Frage wurde dann zwar von BAKIEV et al. (2005) diskutiert, aber nicht abschließend beantwortet; das Konzept wurde weder von MALLOW et al. (2003) noch DE SMEDT (2006) übernommen, die *Vipera nikolskii* als eine valide Art ansahen. Der *Vipera berus*-Komplex wurde von KALYABINA-HAUF et al. (2004) überarbeitet und von NILSON et al. (2005) ausführlich revidiert. Ihre Arrangements finden sich auch hier durch die Anerkennung von vier Unterarten wieder: *Vipera berus berus* (LINNAEUS, 1758), *Vipera berus bosniensis* BOETTGER *in* MOJSISOVICS, 1889, *Vipera berus nikolskii* VEDMEDERJA, GRUBANT & RUDAYEVA, 1986 und *Vipera berus sachalinensis* TSAREWSKY, 1916. JOGER in DAVID & INEICH (1999) gab an, dass das Taxon *bosniensis* möglicherweise als Art Anerkennung finden sollte.

Eine der am wenigsten bekannten Vipern ist *Vipera darevskii* VEDMEDERJA, ORLOV & TUNIYEV, 1986. Die Morphologie, Verbreitung, Beziehungen und der Status dieser seltenen Art wurden von GENIEZ & TEYNIÉ (2005) diskutiert.

Vipera dinniki NIKOLSKY, 1913, ist ein über lange Zeit hinweg mit *Vipera kaznakovi* NIKOLSKY, 1909 gleichgesetztes Taxon, das von ORLOV & TUNIYEV (1986, 1990), NILSON et al. (1994, 1995) und ORLOV & TUNIYEV (2005) besprochen worden ist.

Das Taxon *Vipera ebneri* KNOEPFFLER & SOCHUREK, 1955, wurde als Synonym von *Vipera ursinii eriwanensis* (REUSS, 1933) angesehen (siehe DAVID & INEICH 1999), bis es von NILSON & ANDRÉN (2001) als Art revalidiert wurde; dieses Konzept wurde nachfolgend von MALLOW et al. (2003) anerkannt, doch von JOGER & DELY (2005) wurde *Vipera ebneri* nur als Unterart von *V. renardi* betrachtet.

Vipera eriwanensis (REUSS, 1933) was long regarded as a subspecies of *Vipera ursinii* (BONAPARTE, 1835) sensu lato. It was formally afforded species rank by HÖGGREN et al. (1993), a view followed by NILSON et al. (1994, 1995), NILSON & ANDRÉN (2001), Mallow et al. (2003), and ANANJEVA et al. (2004, 2006).
Vipera lotievi NILSON, TUNIYEV, ORLOV, HÖGGREN & ANDRÉN, 1995, was considered to be a subspecies of *V. renardi* by JOGER & DELY (2005) even though its specific distinctness had earlier been demonstrated by NILSON & ANDRÉN (2001); the views of the latter authors were accepted by MALLOW et al. (2003), ANANJEVA et al. (2004, 2006), and DE SMEDT (2006).
Both *Vipera magnifica* TUNIYEV & OSTROVSKIKH, 2001 and *Vipera orlovi* TUNIYEV & OSTROVSKIKH, 2001 were previously confused with *Vipera kaznakovi*. Clarifying information was then provided by ANANJEVA et al. (2004, 2006).
"*Vipera pontica*" BILLING, NILSON & SATTLER, 1990, has remained a controversial taxon. The species was accepted by GOLAY et al. (1993), DAVID & INEICH (1999), and MALLOW et al. (2003), whereas it was considered a synonym of *Vipera barani* by BARAN et al. (2001) and AVCI et al. (2004). No specimen has ever been collected after the type material. The possibility that the type specimens resulted from a fortuitous hybridization between two other species was briefly mentioned by BILLING et al. (1990). JOGER *in* DAVID & INEICH (1999) added: "Whether *Vipera pontica* is a distinct, valid species or is based on aberrant specimens of *V. ammodytes* [*transcaucasiana*] should be discussed". According to A. TEYNIÉ (pers. comm., December 2007), the hypothesis of hybridization is very likely of substance. *Vipera pontica* may be a hybrid between *Vipera barani*, with which it shares many characters, and either *Vipera kaznakovi* or more probably *Vipera ammodytes transcaucasiana*. As a consequence, we here consider *V. pontica* to be a distinct taxon, but its unresolved status is pointed out by placing it within brackets.

Vipera renardi (CHRISTOPH, 1861) has long been considered a subspecies of *Vipera ursinii* (BONAPARTE, 1835). It was considered a valid species by HÖGGREN et al. (1993), a view accepted by NILSON et al. (1993, 1994, 1995) and NILSON & ANDRÉN (2001). It was also followed by MALLOW et al. (2003), ANANJEVA et al. (2004, 2006) and DE SMEDT (2006). The arrangement published by SZCZERBAK (2003) is out of date. Three subspecies are present in the region covered by this volume: *Vipera renardi renardi* (CHRISTOPH, 1861), *Vipera renardi bashkirovi* BAKIEV, GARANIN, LITVINOV, PAVLOV & RATNIKOV, 2004, and *Vipera renardi tienshanica* NILSON & ANDRÉN, 2001 (see also VOGEL 2006).
Lastly, we here follow entirely the arrangement proposed by NILSON & ANDRÉN (2001) in their revision of the *Vipera ursinii* (BONAPARTE, 1835) complex, which was also accepted by DE SMEDT (2006), but not by DELY & JOGER (2005). As a consequence, five subspecies are accepted here: *Vipera ursinii ursinii* (BONAPARTE, 1835), *Vipera ursinii graeca* NILSON & ANDRÉN, 1988, *Vipera ursinii macrops* MÉHELŸ, 1911, *Vipera ursinii moldavica* NILSON, ANDRÉN & JOGER, 1993, and *Vipera ursinii rakosiensis* MÉHELŸ, 1894.

Vipera eriwanensis (REUSS, 1933) wurde lange Zeit als eine Unterart von *Vipera ursinii* (BONAPARTE, 1835) sensu lato behandelt. HÖGGREN et al. (1993) sprachen ihr dann formell Artrang zu, eine Ansicht, der NILSON et al. (1994, 1995), NILSON & ANDRÉN (2001), MALLOW et al. (2003) und ANANJEVA et al. (2004, 2006) folgten.
Vipera lotievi NILSON, TUNIYEV, ORLOV, HÖGGREN & ANDRÉN, 1995, wurde von JOGER & DELY (2005) als Unterart von *V. renardi* angesehen, obwohl ihr Status als valide Art schon zuvor von NILSON & ANDRÉN (2001) belegt worden war; die Ansicht letztgenannter Autoren wurde von MALLOW et al. (2003), ANANJEVA et al. (2004, 2006) und DE SMEDT (2006) anerkannt.
Sowohl *Vipera magnifica* TUNIYEV & OSTROVSKIKH, 2001, als auch *Vipera orlovi* TUNIYEV & OSTROVSKIKH, 2001, sind in der Vergangenheit mit *Vipera kaznakovi* verwechselt worden. ANANJEVA et al. (2004, 2006) klärte die Situation.
„*Vipera pontica*" BILLING, NILSON & SATTLER, 1990, ist ein umstrittenes Taxon geblieben. Die Art wurde von GOLAY et al. (1993), DAVID & INEICH (1999) und MALLOW et al. (2003) anerkannt, wohingegen BARAN et al. (2001) und AVCI et al. (2004) sie als Synonym von *Vipera barani* ansahen. Nach dem Typenmaterial ist kein weiteres Exemplar mehr gefunden worden. Die Möglichkeit, dass es sich bei den Typen um das Ergebnis einer zufälligen Hybridisierung zweier anderer Arten handelt, wurde von BILLING et al. (1990) kurz erwähnt. JOGER in DAVID & INEICH (1999) fügte dem hinzu: „Ob es sich bei *Vipera pontica* um eine eigenständige, valide Art handelt oder aber um aberrante Exemplare von *V. ammodytes* [*transcaucasiana*] muss erst noch untersucht werden". A. TEYNIÉ (pers. Mitt., Dezember 2007) zufolge hat die Hypothese einer Hybridisierung sehr wahrscheinlich Substanz. *Vipera pontica* könnte ein Kreuzungsprodukt zwischen *Vipera barani* (mit der sie zahlreiche Merkmale teilt) und entweder *Vipera kaznakovi* oder, was wahrscheinlicher ist, *Vipera ammodytes transcaucasiana* sein. An dieser Stelle betrachten wir *V. pontica* als eigenständiges Taxon, unterstreichen allerdings die problematische Stellung dieses Taxons dadurch, dass wir es in Klammern führen.
Vipera renardi (CHRISTOPH, 1861) wurde über lange Zeit als Unterart von *Vipera ursinii* (BONAPARTE, 1835) betrachtet. HÖGGREN et al. (1993) sprachen ihr dann den Rang einer eigenen Art zu, eine Ansicht, die von NILSON et al. (1993, 1994, 1995) und NILSON & ANDRÉN (2001) geteilt wurde. Sie wurde dann auch von MALLOW et al. (2003), ANANJEVA et al. (2004, 2006) und DE SMEDT (2006) übernommen. Die von SZCZERBAK (2003) beibehaltene Einteilung ist veraltet. In der hier behandelten Region sind drei Unterarten vertreten: *Vipera renardi renardi* (CHRISTOPH, 1861), *Vipera renardi bashkirovi* BAKIEV, GARANIN, LITVINOV, PAVLOV & RATNIKOV, 2004 und *Vipera renardi tienshanica* NILSON & ANDRÉN, 2001 (siehe auch VOGEL 2006).
Schließlich folgen wir hier der von NILSON & ANDRÉN (2001) in ihrer Revision des *Vipera ursinii* (BONAPARTE, 1835) Komplexes vorgeschlagenen Gliederung, die auch von DE SMEDT (2006), aber nicht von DELY & JOGER (2005) übernommen wurde. Folglich sehen wir hier fünf Unterarten als gültig an: *Vipera ursinii ursinii* (BONAPARTE, 1835), *Vipera ursinii graeca* NILSON & ANDRÉN, 1988, *Vipera ursinii macrops* MÉHELŸ, 1911, *Vipera ursinii moldavica* NILSON, ANDRÉN & JOGER, 1993 und *Vipera ursinii rakosiensis* MÉHELŸ, 1894.

Distribution maps

The maps included in this book are meant to summarize the approximate geographical ranges of all species and subspecies considered in the present volume. Maps were drawn as accurately as possible from data available in the literature. However, these maps take into account only the geographical ranges and neither altitudinal limits nor habitat characteristics. Any given range may therefore include areas that are not actually habitable to the respective taxon. Three main categories of publications were used: taxonomic works (systematic revisions, taxonomic reviews, and original descriptions), regional faunae and atlases, and the general atlases of GASC et al. (1999) for European countries and LEVITON et al. (1992) for the Near and Middle East, respectively.

Verbreitungskarten

Die in diesem Buch enthaltenen Verbreitungskarten sollen die ungefähre geographische Verbreitung aller im vorliegenden Band abgehandelten Arten und Unterarten zusammenfassend darstellen. Diese Karten wurden nach den in der Literatur enthaltenen Angaben so genau wie möglich erstellt. Dennoch wird hier jeweils lediglich die geographische Ausbreitung berücksichtigt, nicht jedoch Einschränkungen durch Höhenlagen oder Biotopansprüche. Die Verbreitungsangaben können daher in jedem Fall Bereiche einschließen, die für das entsprechende Taxon tatsächlich nicht bewohnbar sind. In diesem Zusammenhang wurden drei Sorten von Veröffentlichungen herangezogen: taxonomische Arbeiten (systematische Revisionen, taxonomische Überarbeitungen und Erstbeschreibungen), Übersichten und Atlanten zu Regionalfaunen und die allgemeinen Atlanten von GASC et al. (1999) für die Länder Europas bzw. von LEVITON et al. (1992) für den Nahen und Mittleren Osten.

Pictograms

Biological data are provided by series of pictograms, with each series dealing with a particular category of biological and ecological characters. This information also originates predominantly from the literature. The pictograms are detailed on a fold-out page on the last page of the book.

These symbols should only be regarded as generalized summaries of known ecological information. In some instances, such as diet and circadian activity, we indicate the most common behavioural patterns. A species may be diurnal in spring and become nocturnal during the hot summer months. We attempted to define as accurately as possible the various habitats found in the region covered by the present volume, which range from the cool boreal forests to some of the hottest deserts in the world.

Maximum lengths are given with reference to the largest documented specimen; these sizes are therefore not the commonly reached maximum lengths. We ignored some unsubstantiated statements of gigantism. We also relied mostly upon data referring to wild-caught specimens. Some published record sizes refer to captive-bred snakes, which may have attained their large size as a function of their artificial environment.

Most of the ecological information given can be sourced in the references quoted. We also investigated certain populations, but to save space for pictures of living animals, we refrained to list the relevant details.

Piktogramme

Biologische Informationen werden jeweils durch eine Reihe von Piktogrammen zusammengefasst, wobei sich jede Reihe auf eine spezifische Kategorie von biologischen und ökologischen Merkmalen bezieht. Diese Informationen entstammen ebenfalls hauptsächlich der Literatur. Die einzelnen Piktogramme sind nach Seite 160 erklärt.

Die verwendeten Symbole sollten lediglich als verallgemeinerte Zusammenfassung bekannt gewordener ökologischer Erkenntnisse verstanden werden. In manchen Fällen, wie etwa bei Nahrung und Aktivitätszeitraum, beschränken wir uns auf die hauptsächlichen Verhaltensmuster. Eine Art kann somit durchaus im Frühjahr tagaktiv sein, während der heißen Sommermonate aber zu Nachtaktivität übergehen. Wir haben auch versucht, die im vorliegenden Band anzutreffenden verschiedenen Lebensräume so genau wie möglich darzustellen. Diese reichen von den kühlen Wäldern der Borealis zu einigen der heißesten Wüsten der Welt.

Die angegebenen Maximallängen beziehen sich jeweils auf das größte dokumentierte Exemplar; es handelt sich somit nicht um die in der Regel erreichte Größe. Wir haben dabei einige unbelegte Angaben zu Riesenwüchsigkeit übergangen und uns in der Hauptsache auf Angaben zu freilebenden Exemplaren beschränkt. Einige veröffentlichte Rekordgrößen beziehen sich auf in Gefangenschaft aufgezogene Tiere, die ihre Größe möglicherweise aufgrund ihrer künstlichen Lebensumstände erreicht haben mögen.

Die meisten der ökologischen Informationen können in den angegebenen Literaturquellen nachgelesen werden. Wir haben darüber hinaus bestimmte Populationen selbst untersucht, jedoch macht es der hier zur Verfügung stehende Platz unmöglich, diese im Detail aufzulisten.

References/Literaturverzeichnis

Adnagulov, E.V., I.G. Tarasov & V.V. Gorobeiko (2000): New data on Amphibians and Reptiles distribution in the Russian Far East. – Russian Journal of Herpetology, **7** (2): 139–154.

Ananjeva, N.B., N.L. Orlov, R.G. Khalikov, I.S. Darevsky, S.A. Ryabov & A.V. Barabanov (2004): Colored atlas of the reptiles of the North Eurasia (Taxonomic diversity, distribution, conservation status), St. Petersburg, Zoological Institute: 1–230. (In Russian).

Ananjeva, N.B., N.L. Orlov, R.G. Khalikov, I.S. Darevsky, S.A. Ryabov & A.V. Barabanov (2006): The Reptiles of Northern Eurasia. Taxonomic diversity, distribution, conservation status, Sofia, Pensoft Publishers: 1–245.

Arnold, E.N., M.D. Robinson & S. Carranza (2009): A preliminary analysis of phylogenetic relationships and biogeography of the dangerously venomous Carpet Vipers, *Echis* (Squamata, Serpentes, Viperidae) based on mitochondrial DNA sequences. – Amphibia-Reptilia, **30**: 273–282.

Auffenberg, W. & H. Rehman (1991): Studies on Pakistan reptiles. Part 1. The genus *Echis* (Viperidae). – Bulletin of the Florida Museum of Natural History, (Biological Sciences), **35** (5): 263–314.

Avci, A., N. Üzum & K. Olgun (2004): A new record of *Vipera barani* Böhme et Joger, 1983 (Reptilia, Viperidae) from North-Eastern Anatolia, Turkey. – Russian Journal of Herpetology, **11** (1): 77–79.

Babocsay, G. (2003): Geographic variation in *Echis coloratus* (Viperidae, Ophidia) in the Levant with the description of a new subspecies. – Zoology in the Middle East, 29: 13–32.

Babocsay, G. (2004): A new species of saw-scaled viper of the *Echis coloratus* complex (Ophidia: Viperidae) from Oman, Eastern Arabia. – Systematics and Biodiversity, **1** [2003] (4): 503–514.

Bakiev, A., W. Böhme & U. Joger (2005): *Vipera* (*Pelias*) [*berus*] *nikolskii* Vedmederya, Grubant & Rudaeva, 1986 – Waldsteppenotter. *In*: Joger, U. & N. Stümpel (Ed.): Handbuch der Reptilien und Amphibien Europas. Band 3/IIB. Schlangen (Serpentes) III Viperidae, Wiebelsheim (Germany), AULA-Verlag GmbH: 293–309.

Baran, I., U. Joger, B. Kutrup & O. Turkozan (2001): On new specimens of *Vipera barani* Böhme & Joger, 1983, from northeastern Anatolia, and implications for the validity of *Vipera pontica* Billing, Nilson & Sattler, 1990 (Reptilia, Viperidae). – Zoology in the Middle East, **23**: 47–53.

Biella, H.-J. & E. Blätter (1992): Zur geographischen Variation eidonomischer Merkmale bei der Sandviper (*Vipera ammodytes*). Mitteilungen aus dem Zoologischen Museum in Berlin, **68** (2): 249–258.

Bostanchi, H., S.C. Anderson, H.G. Kami & T.J. Papenfuss (2006): A new species of *Pseudocerastes* with elaborate tail ornamentation from western Iran (Squamata: Viperidae). – Proceedings of the California Academy of Sciences, (4th Ser.), **57** (14): 443–450.

Bour, R. (1993): Les voyages de Peter Simon Pallas et l'origine de *Coluber halys* (Serpentes Viperidae). Bulletin mensuel de la Société Linnéenne de Lyon, **62** (9): 328–340.

Cherlin, V.A. (1990): Taxonomic revision of the snake genus *Echis* (Viperidae). II. An analysis of taxonomy and description of new forms. *In*: Borkin, L.J. (Ed.), Reptiles of mountain and arid territories: systematics and distribution, Trudy Zoologicheskogo Instituta Akademii Nauk. SSSR, Leningrad, **207**: 193–223. (In Russian, English summary).

Cherlin, V.A. & L.J. Borkin (1990): Taxonomic revision of the snake genus *Echis* (Viperidae). I. Analysis of the history of study and synonymy. *In*: Borkin, L.J. (Ed.), Reptiles of mountain and arid territories: systematics and distribution, Trudy Zoologicheskogo Instituta Akademii Nauk. SSSR, Leningrad, **207**: 175–192. (In Russian, English summary).

Chikin, Y. & N.N. Szczerbak (1992): New subspecies of bluntnosed viper, *Vipera lebetina cernovi*, ssp. nov. (Reptilia, Viperidae) from Middle Asia. – Vestnik Zoologiy, **1992** (6): 45–49. (In Russian, English summary).

David, P., N.B. Ananjeva, I. Das & M. Golubev (1999): Translation of the original description of *Vipera obtusa* Dwigubsky, 1832 (Serpentes: Viperidae: Viperinae), with designation of a neotype. – Russian Journal of Herpetology, **6** (3): 193–198.

David, P. & I. Ineich (1999): Les Serpents venimeux du monde: sytématique et répartition. – Dumerilia, Paris, **3**: 1–499.

Dely, O.G. & U. Joger (2005): *Vipera* (*Pelias*) *ursinii* Bonaparte, 1935 – Wiesenotter. *In*: Joger, U. & N. Stümpel (Ed.): Handbuch der Reptilien und Amphibien Europas. Band 3/IIB. Schlangen (Serpentes) III Viperidae, Wiebelsheim (Germany), AULAVerlag GmbH: 375–414.

De Smedt, J. (2006): The Vipers of Europe, Halblecht (Germany), JDS Verlag: i–v + 1–340.

Dirksen, L. & W. Böhme (2005): *Gloydius halys* (Pallas, 1776) – Halysotter. *In*: Joger, U. & N. Stümpel (Ed.): Handbuch der Reptilien und Amphibien Europas. Band 3/IIB. Schlangen (Serpentes) III Viperidae, Wiebelsheim (Germany), AULA-Verlag GmbH: 7–22.

Dobiey, M. & G. Vogel (2007): Terralog Vol. 15. Venomous snakes of Africa – Giftschlangen Afrika, Frankfurt am Main, Edition Chimaira & Rodgau, Aqualog Verlag ACS: 1–148.

Egan, D. (2007): Snakes of Arabia. A Field Guide to the Snakes of the Arabian Peninsula and its Shores, Dubai, Motivate Publishing: 1–208.

Garrigues, T. C. Dauga, E. Ferquel, V. Choumet & A.-B. Failloux (2005): Molecular phylogeny of *Vipera* Laurenti, 1768 and the related genera *Macrovipera* (Reuss, 1927) and *Daboia* (Gray, 1842), with comments about neurotoxic *Vipera aspis aspis* populations. – Molecular Phylogenetics and Evolution, **35**: 35–47.

Geniez, P. & A. Teynié (2005): Discovery of a population of the critically endangered *Vipera darevskii* Vedmederja, Orlov & Tuniyev, 1986 in Turkey, with new elements on its identification. – Herpetozoa, **18** (1/2): 25–33.

Gloyd, H.K. & R. Conant (1990): Snakes of the *Agkistrodon* complex: a monographic review. Athens (Ohio), Society for the Study of Amphibians and Reptiles, Contributions to Herpetology, **6**: i–vi + 1–614, Pls. 1–52.

Golay, P., H.M. Smith, D.G. Broadley, J.R. Dixon, C.J. McCarthy, J.-C. Rage, B. Schätti & M. Toriba (1993): *Endoglyphs and other major venomous snakes of the world. A checklist*, Aïre-Genève, Azemiops S.A. Herpetological Data Center: i–xv + 1–478.

Gumprecht, A., F. Tillack, N.L. Orlov, A. Captain & S. Ryabov (2004): Asian Pitvipers, Berlin, Geitje Books: 1–368.

Heckes, U., H.-J. Gruber & N. Stümpel (2005): *Vipera* (*Vipera*) *ammodytes* (Linnaeus, 1758) – Hornotter, Sandviper. *In*: Joger, U. & N. Stümpel (Ed.): Handbuch der Reptilien und Amphibien Europas. Band 3/IIB. Schlangen (Serpentes) III Viperidae, Wiebelsheim (Germany), AULA-Verlag GmbH: 81–150.

Herrmann, H.-W. & U. Joger (1996 [dated 1997]): Evolution of viperine snakes. *In*: Thorpe, R. S., W. Wüster & A. Malhotra (Ed.), Venomous snakes. Ecology, evolution and snakebite, Oxford,

Clarendon Press & London, The Zoological Society of London, Symposia of the Zoological Society of London, **70**: 43–61.

Herrmann, H.-W., U. Joger & G. Nilson (1992): Phylogeny and systematics of viperine snakes. III: Resurrection of the genus *Macrovipera* (Reuss, 1927) as suggested by biochemical evidence. – Amphibia-Reptilia, **13**: 375–383.

Höggren, M., G. Nilson, C. Andrén & N.L. Orlov (1993): Vipers of the Caucasus: natural history and systematic review. – Herpetological natural History, **1** (2): 11–19.

Joger, U. (2005): *Montivipera* Nilson, Tuniyev, Andrén, Orlov, Joger & Herrmann, 1999. *In*: Joger, U. & N. Stümpel (Ed.): Handbuch der Reptilien und Amphibien Europas. Band 3/IIB. Schlangen (Serpentes) III Viperidae, Wiebelsheim (Germany), AULA-Verlag GmbH: 61–62.

Joger, U. & O.G. Dely (2005): *Vipera (Pelias) renardi* (Christoph, 1861) – Steppenotter. *In*: Joger, U. & N. Stümpel (Ed.): Handbuch der Reptilien und Amphibien Europas. Band 3/IIB. Schlangen (Serpentes) III Viperidae, Wiebelsheim (Germany), AULA-Verlag GmbH: 343–354.

Joger, U., P. Lenk, I. Baran, W. Böhme, T. Ziegler, P. Heidrich, & M. Wink (1995): The phylogenetic position of *Vipera barani* and *V. nikolskii* within the *Vipera berus* complex. *In*: Anonymous, Programme and abstracts – Societas Europaea Herpetologica. 8th Ordinary general meeting, Bonn, Germany, 23–27 August 1995, Bonn, Museum Alexander Koenig: 68–69.

Joger, U., P. Lenk, I. Baran, W. Böhme, T. Ziegler, P. Heidrich, & M. Wink (1997): The phylogenetic position of *Vipera barani* and *V. nikolskii* within the *Vipera berus* complex. *In*: Böhme, W., W. Bischoff & T. Ziegler (Ed.), Herpetologia Bonnensis (Proceedings of the 8th Ordinary general meeting of SEH, Bonn, 1995), Bonn, Societas Europaea Herpetologica: 185–194.

Joger, U. & N. Stümpel (Ed.) (2005): Handbuch der Reptilien und Amphibien Europas. Band 3/IIB. Schlangen (Serpentes) III Viperidae, Wiebelsheim (Germany), AULA-Verlag GmbH: i–x + 1–420.

Kalyabina-Hauf, S., S. Schweiger, U. Joger, W. Mayer, N. Orlov & M. Wink (2004): Phylogeny and systematics of adders (*Vipera berus* complex). – Mertensiella, **15**: 7–16.

Khan, M.S. (1985): Taxonomic note on *Bungarus caeruleus* and *B. sindanus* Boulenger. – The Snake, **17** (1): 71–78.

Kuch, U. (2004): *Bungarus sindanus* Boulenger, 1897, an addition to the venomous snake fauna of Afghanistan. – Herpetozoa, **16** (3/4): 171–173.

Lenk, P., S. Kalyabina, M. Wink & U. Joger (2001): Evolutionary relationships among the True Vipers (Reptilia: Viperidae) inferred from mitochondrial DNA sequences. – Molecular Phylogenetics and Evolution, **19**(1): 94–104.

Mallow, D., D. Ludwig & G. Nilson (2003): True Vipers. Natural history and toxinology of Old World Vipers, Malabar (Florida), Krieger Publishing Company: i–xiv + 1–359.

Nilson, G. & C. Andrén (1986): The mountain vipers of the Middle East – the *Vipera xanthina* complex (Reptilia, Viperidae). – Bonner zoologische Beiträge, **20**: 1–90.

Nilson, G. & C. Andrén (1988a): A new subspecies of the subalpine meadow viper, *Vipera ursinii* (Bonaparte) (Reptilia, Viperidae), from Greece. – Zoologica Scripta, **17** (3): 311–314.

Nilson, G. & C. Andrén (1988b): *Vipera lebetina transmediterranea*, new subspecies of Viper from North Africa, with remarks on the taxonomy of *Vipera lebetina* and *Vipera mauritanica* (Reptilia: Viperidae). – Bonner zoologische Beiträge, **39** (4): 371–379.

Nilson, G. & C. Andrén (1992): The species concept in the *Vipera xanthina* complex: reflecting evolutionary history or hiding biological diversity. – Amphibia-Reptilia, **13** (4): 421–424.

Nilson, G. & C. Andrén (1996 [dated 1997]): Evolution, systematics and biogeography of Palaearctic vipers. *In*: Thorpe, R.S., W. Wüster & A. Malhotra (Ed.), Venomous snakes. Ecology, evolution and snakebite, Oxford, Clarendon Press & London, The Zoological Society of London, Symposia of the Zoological Society of London, **70**: 31–42.

Nilson, G. & C. Andrén (2001): The meadow and steppe vipers of Europe and Asia – The *Vipera (Acridophaga) ursinii* complex. Acta Zoologica Academiae Scientiarum Hungaricae, **47** (2–3): 87–267.

Nilson, G., C. Andrén & U. Joger (1993): A re-evaluation of the taxonomic status of the Moldavian steppe viper based on immunological investigations, with discussion of the hypothesis of secondary intergradation between *Vipera ursinii rakosiensis* and *Vipera (ursinii) renardi*. – Amphibia-Reptilia, **14** (1): 45–57.

Nilson, G., C. Andrén & W. Völki (2005): *Vipera (Pelias) berus* (Linnaeus, 1758) – Kreuzotter. *In*: Joger, U. & N. Stümpel (Ed.): Handbuch der Reptilien und Amphibien Europas. Band 3/IIB. Schlangen (Serpentes) III Viperidae, Wiebelsheim (Germany), AULA-Verlag GmbH: 213–292.

Nilson, G., M. Höggren, B.S. Tuniyev, N.L. Orlov & C. Andrén (1994): Phylogeny of the vipers of the Caucasus (Reptilia, Viperidae). – Zoologica Scripta, **23** (4): 353–360.

Nilson, G. & N. Rastegar-Pouyani (2007): *Walterinnesia aegyptia* Lataste, 1887 (Ophidia: Elapidae) and the status of *Naja morgani* Mocquard 1905. – Russian Journal of Herpetology, **14** (1): 7–14.

Nilson, G., B. Tuniyev, C. Andrén, N. Orlov, U. Joger & H.W. Herrmann (1999): Taxonomic position of the *Vipera xanthina* complex. – Kaupia, **8**: 99–102.

Nilson, G., B. Tuniyev, N.L. Orlov, M. Höggren & C. Andrén, C. (1995): Systematics of the vipers of the Caucasus: polymorphism or sibling species? – Asiatic herpetological Researches, **6**: 1–26.

Obst, F.J. (1983): Zur Kenntnis der Schlangengattung *Vipera* (Reptilia, Serpentes, Viperidae). – Zoologische Abhandlungen, Staatliches Museum für Tierkunde in Dresden, **38** (13): 229–235.

Orlov, N.L. & A.V. Barabanov (2000a): Analysis of nomenclature, classification, and distribution of the *Agkistrodon halys-Agkistrodon intermedius* complexes: a critical review. – Russian Journal of Herpetology, **6** (3): 167–192.

Orlov, N.L. & A.V. Barabanov (2000b): About type localities for some species of the genus *Gloydius* Hoge et Romano-Hoge, 1981 (Crotalinae: Viperidae: Serpentes). – Russian Journal of Herpetology, **7** (2): 159–160.

Orlov, N.L. & B.S. Tuniyev (1986): The recent areas, their possible genesis and the phylogeny of three viper species of Eurosiberian group of the *Vipera kaznakowi* complex in the Caucasus. – Trudy Zoologicheskogo Instituta Akademii Nauk. SSSR, Leningrad, **157**: 107–135. (In Russian, English summary).

Orlov, N.L. & B.S. Tuniyev (1990): Three species in the *Vipera kaznakowi* complex (Eurosiberian group) in the Caucasus: their present distribution, possible genesis and phylogeny. Asiatic herpetological Research, **3**: 1–36.

Orlov, N.L. & B.S. Tuniyev (2005): *Vipera (Pelias) dinniki* Nikolskij, 1913 – Dinniks Kaukasusotter. *In*: Joger, U. & N. Stümpel (Ed.): Handbuch der Reptilien und Amphibien Europas. Band 3/IIB. Schlangen (Serpentes) III Viperidae, Wiebelsheim (Germany), AULA-Verlag GmbH: 310–323.

PALLAS, P.S. (1776): Reise durch Verschiedene Provinzen des Russischen Reichs. Dritter Theil. Vom Jahr 1772 und 1773, St. Petersburg, Kayserlichen Academie der Wissenschaften: 1–760, i–xxxii, Pls. 8–9, A-Z, Aa-Nn.

PALLAS, P.S. (1814): Zoographia Rosso-Asiatica, sistens omnium animalium extenso Imperio Rossico et adjacentibus maribus observatorum recensionem, domicilia, mores et descriptiones, anatomen atque icones plurimorum. Vol. III. Animalia Monocardia seu frigidi sanguinis Imperii Rosso-Asiatici, Petropoli [St. Petersburg], Officina Caes. Academiae Scientiarum: 1–427.

POOKE, C.E., U. JOGER, N. STÜMPEL, C.J. MCCARTHY & W. WÜSTER (2009): When continents collide: Phylogeny, historical biogeography and systematics of the medically important viper genus *Echis* (Squamata: Serpentes: Viperidae). – Molecular Phylogenetics and Evolution, **53**: 792–807.

SAINT GIRONS, H. (1978): Morphologie externe comparée et systématique des Vipères d'Europe (Reptilia, Viperidae). – Revue suisse de Zoologie, **85** (3): 565–595, Pls. 1–2.

SAINT GIRONS, H., R. DUGUY & J. DETRAIT (1983): Les vipères du sud du Massif Central. – Bulletin de la Société d'Histoire Naturelle de Toulouse, **119**: 81–86.

SCHÄTTI, B. (2001): A new species of *Coluber* (sensu lato) from the Dahlak Islands, Eritrea, with a review of the herpetofauna of the archipelago. Russian Journal of Herpetology, **8** (2): 139–148.

SCHÄTTI, B., I. BARAN & H. SIGG (1991): Rediscovery of the Bolkar viper: morphological variation and systematic implications on the '*Vipera xanthina* complex'. – Amphibia-Reptilia, **12** (3): 305–327.

SCHÄTTI, B., I. BARAN & H. SIGG (1992): The '*Vipera xanthina* complex' – a reply to NILSON and ANDRÉN. – Amphibia-Reptilia, **13** (4): 425.

SCHÄTTI, B. & J. GASPERETTI (1994): A contribution to the herpetofauna of Southwest Arabia. – Fauna Saudi Arabia, **14**: 348–423.

SCHWEIGER, M. (1995): Morphological variation in *Vipera raddei* from the area of the former USSR. *In*: Anonymous, Programme and abstracts – 8th Ordinary general meeting, Societas Europaea Herpetologica, Bonn, Germany, 23–27 August 1995, Bonn, Museum Alexander Koenig: 99.

STRAUCH, A.A. (1873): Die Schlangen des Russischen Reichs, in systematischer und zoogeographischer Beziehung. – Mémoires de l'Académie Impériale des Sciences de St. Pétersbourg, (7), **21** (4): 1–287, Pls. 1–6.

STÜMPEL, N. & U. JOGER (2009): Recent advances in phylogeny and taxonomy of Near and Middle Eastern Vipers – an update. *In*: NEUBERT, E., Z. AMR, S. TAITI & B. GÜMÜS (Eds), Animal Biodiversity in the Middle East. Proceedings of the First Middle Eastern Biodiversity Congress, Aqaba, Jordan, 20–23 October 2008. ZooKeys, **31**: 179–191.

SZCZERBAK, N.N. (2003): Guide to the Reptiles of the Eastern Palearctic, Malabar (Florida), Krieger Publishing Company: xvii + 1–260.

TORIBA, M. (1986): The classification of the genus *Agkistrodon* with 21 scale rows in Eastern Asia. – Acta Herpetologica Sinica, (New series), **5** (1): 61–62.

TRAPE, J.-F., L. CHIRIO, D.G. BROADLEY & W. WÜSTER (2009): Phylogeography and systematic revision of the Egyptian cobra (Serpentes: Elapidae: *Naja haje*) species complex, with the description of a new species from West Africa. – Zootaxa, **2236**: 1–25.

TRAPE, J.-F., Y. MANÉ & I. INEICH (2006): *Atractaspis microlepidota*, *A. micropholis* et *A. watsoni* en Afrique occidentale et centrale. – Bulletin de la Société Herpétologique de France, **119**: 5–16.

TRUTNAU, L., W. BÖHME & U. JOGER (2005): *Vipera* (*Vipera*) *aspis* (LINNAEUS, 1758) – Aspisviper. *In*: JOGER, U. & N. STÜMPEL (Ed.): Handbuch der Reptilien und Amphibien Europas. Band 3/IIB. Schlangen (Serpentes) III Viperidae, Wiebelsheim (Germany), AULA-Verlag GmbH: 151–185.

URSENBACHER, S., A. CONELLI, P. GOLAY, J.-C. MONNEY, M.A.L. ZUFFI, G. THIERY, T. DURAND & L. FUMAGALLI (2006): Phylogeography of the asp viper (*Vipera aspis*) inferred from mitochondrial DNA sequence data: Evidence for multiple Mediterranean refugial areas. – Molecular Phylogenetics and Evolution, **38**: 546–552.

URSENBACHER, S., S. SCHWEIGER, L. TOMOVIC, L., J. CRNOBRNJA-ISAILOVIC, L. FUMAGALLI & W. MAYER (2008): Molecular phylogeography of the nose-horned viper (*Vipera ammodytes* (LINNAEUS, 1758)): evidence for high genetic diversity and multiple refugia in the Balkan peninsula. – Molecular Phylogenetics and Evolution, , 46: 1116–1128

VIDAL, N., W.R. BRANCH, O.S.G. PAUWELS, S.B. HEDGES, D.G. BROADLEY, M. WINK, C. CRUAUD, U. JOGER & Z.T. NAGY (2008): Dissecting the major African snake radiation: a molecular phylogeny of the Lamprophiidae FITZINGER (Serpentes, Caenophidia). – Zootaxa, **1945**: 51–66.

VIDAL, N., A.-S. DELMAS, P. DAVID, C. CRUAUD, A. COULOUX & S.B. HEDGES (2007): The phylogeny and classification of caenophidian snakes inferred from seven nuclear protein-coding genes – Comptes Rendus Biologies, **330**: 182–187.

VOGEL, G. (2006): Terralog Vol. 14. Venomous snakes of Asia - Giftschlangen Asiens. Frankfurt am Main, Edition Chimaira & Rodgau, Aqualog Verlag ACS: 1–148.

WALLACH, W., W. WÜSTER & D.G. BROADLEY (2009): In praise of subgenera: taxonomic status of cobras of the genus *Naja* LAURENTI (Serpentes: Elapidae). – Zootaxa, **2236**: 26–36.

WERNER, Y.L., A. LE VERDIER, D. ROSENMAN & N. SIVAN (1991): Systematics and zoogeography of *Cerastes* (Ophidia: Viperidae) in the Levant: I. Distinguishing Arabian from African "*Cerastes cerastes*". – The Snake, **23**: 90–100.

WERNER, Y.L. & N. SIVAN (1992): Systematics and zoogeography of *Cerastes* (Ophidia: Viperidae) in the Levant: 2. Taxonomy, ecology, and zoogeography. – The Snake, **24**: 34–49.

WERNER, Y. L., N. SIVAN, V. KUSHNIR & U. MOTRO (1999): A statistical approach to variation in *Cerastes* (Ophidia: Viperidae), with the description of two endemic subspecies. – Kaupia, **8**: 83–97.

WÜSTER, W. (1996): Taxonomic changes and toxinology: systematic revisions of the Asiatic cobras (*Naja naja* species complex). – Toxicon, **34** (4): 399–406.

WÜSTER, W., L. PEPPIN, C. E. POOK & D.E. WALKER (2008): A nesting of vipers: phylogeny and historical biogeography of the Viperidae (Squamata: Serpentes). – Molecular Phylogenetics and Evolution, **49**: 445–459.

WÜSTER, W. & R.S. THORPE (1991): Asiatic cobras: systematics and snakebite. – Experientia, **47**: 205–209.

WÜSTER, W. & R.S. THORPE (1992a): Asiatic cobras: population systematics of the *Naja naja* species complex (Serpentes: Elapidae) in India and Central Asia. – Herpetologica, **48** (1): 69–85.

WÜSTER, W. & R.S. THORPE (1992b): Dentitional phenomena in cobras revisited: spitting and fang structure in the Asiatic species of *Naja* (Serpentes: Elapidae). – Herpetologica, **48** (4): 424–434.

ZUFFI, M., 2002. A critique of the systematic position of the aspic viper subspecies *Vipera aspis aspis* (LINNÉ, 1758), *Vipera aspis atra* MEISNER, 1820, *Vipera aspis francisciredi* LAURENTI, 1768, *Vipera aspis hugyi* SCHINZ, 1833 and *Vipera aspis zinnikeri* KRAMER, 1958. – Amphibia-Reptilia, **23** (2): 191–213.

Acknowledgments/Danksagung

This book could not have been published without the enormous help of photographers from around the world. About 86 professional and amateur herpetologists and naturalists from 25 different countries kindly provided us the outstanding pictures without which the preparation of this book could not have been possible. We received more than 2,500 pictures, of which only a fraction could be presented in this volume after a tough choice. We want to thank all these photographers for their trustfulness and their generosity. We hope they are satisfied with the result. Their amazing pictures speak themselves for them. All these photographers are listed by countries in which they live or work.
Dieses Buch wäre nicht möglich gewesen ohne die großzügige Hilfe zahlreicher Fotografen aus der ganzen Welt. Wir freuen uns, in diesem Band Fotos von 86 professionellen- und Amateur-Wissenschaftlern sowie Naturfreunden aus 25 Ländern zeigen zu können, welche uns ihre teilweise herausragenden Aufnahmen zur Verfügung gestellt haben. Insgesamt haben wir mehr als 2500 Bilder erhalten, wovon hier, nach einer zähen Auswahl, nur ein Bruchteil gezeigt werden kann. Bei allen Leuten, die uns Bilder zur Verfügung gestellt haben, möchten wir uns für ihre Großzügigkeit und ihr Vertrauen bedanken. Wir hoffen, dass Sie mit dem vorliegenden Ergebnis zufrieden sind. Die großartigen Bilder sprechen für sich selbst. Die Fotografen sind im Folgenden nach Ländern, in denen sie leben bzw. arbeiten, zusammengefasst.

Armenia: N. Ayvazian • **Austria:** T. Bader, N. Beigi, T. Böhm, J. Hill, C. Riegler, M. Schweiger • **Belgium:** van der Voort • **Croatia:** B. Horvatic, V. Peranic • **Czech Republic:** V. Gvoždik, D. Hegner, J. Hejduk, D. Jablonski, O. Knobloch, T. Mazuch, D. Modrý, P. Nečas, K. Rozinek, Jan Ševčík • **Egypt:** P. Gasperetti † • **Federation of Russia:** O. V. Korsun, A. Pavlov • **Finland:** K. Lipponen, T. Paasikunnas • **France:** P. Geniez, J. Maran, A. Teynié • **Germany:** F. Baier, D. Bohle, Matthias Dehling, Max Dehling, M. Dobiey, F. Finke, F. Hulbert, K. Kabisch, O. Klingner, A. Koch, G. Kreiner, N. Lutzmann, A. Pieh, F. Siegenthaler, J. de Smedt, B. Trapp, L. Trutnau, G. Vogel, T. Wilms • **Hungary:** G. Babocsay • **India:** V. Deepak, V. K. Mistry, M. Selvan • **Iran:** B. Fathinia, B. Jafari, H. Kami, M. Rajabizadeh • **Israel:** A. Bar, A. Meier • **Kingdom of Saudi Arabia:** A. Abuzeid • **Slovakia:** D. Jandzik • **Spain:** D. Donaire • **Sweden:** L. A. Darafeyenka, M. Karlsson, M. Stahre, L. Westrin • **Switzerland:** H. Billing, P. Brodmann †, S. Dummermuth, K. Mebert, A. Meyer, T. Ott, H. Sigg, H. Wilhelm • **The Netherlands**: E. Evan, P. Janssen, H. van der Ploeg, J. C. Vermeer, E. Wassink • **Turkey:** A. Avci, Y. Kumluta • **Ukraine:** N. M. Beskaravainyi, O. V. Kukushkin • **United Arab Emirates:** X. Eichaker • **United Kingdom:** W. Wüster • **USA:** M. S. Khan, A. E. Leviton, John H. Tashjian

Beside the photographers, several other persons helped us in making contacts or searching for pictures. It is a pleasure to thank:
Neben den Fotografen haben uns weitere Personen entweder bei der Bildersuche oder bei der Suche nach den Adressen möglicher Bildquellen unterstützt:
R. Bour, N. Brachtel, P. Brodmann jun., M. Dobiey, Mrs. P. Gasperetti, P. Hien, Dr. Kasparek, O. Klingner, L. Kratochvil, N. Lutzmann, Prof. J. Meier, T. Monigatti, P. Nečas, V. Peranic, H. van der Ploeg, S. Schweiger, B. Shacham, A. Teynié, D. Trautmann, Y. L. Werner, H. Wilhelm, T. Wilms, and W. Wüster.

For several sections, specialists contributed with their knowledge to avoid mistakes or misidentifications. We are also grateful to:
Bei einigen Abschnitten haben wir uns der Kenntnisse führender Experten für die jeweiligen Gattungen oder Artgruppen bedient, um Fehler oder Fehlbestimmungen zu vermeiden. Wir bedanken uns für die Beratung bei:
Thomas Garrigues (*Vipera aspis* complex), Oliver Klingner (*Vipera ursinii* complex), Ivo Peranic (*Vipera berus* group), Falk Siegenthaler (genus *Macrovipera*), and Wolfgang Wüster (genus *Echis*).

The following persons helped us in the one or other way:
Folgende Personen unterstützen uns auf unterschiedlichste Weise in der einen oder anderen Phase der Manuskripterstellung: T. Dujsebayeva, P. Freed, H. Hellemar, N. Lutzmann, D. V. Malakhov, B. Sterijovski, S. R. Zamfirescu.

Our special thanks are going to two persons: To Mrs. Patricia Gasperetti, who was so kind to permit us to use pictures of her late husband, who died in 2001 and whose contributions are still substantial for the Arabian herpetofauna, and to Peter Brodman Jr. who was so kind to help us to obtain pictures of his father who was both an accomplished naturalist and an amazing photographer. He provided us with the first unforgettable book on the vipers of Europe, which was far ahead of its time and which is still a milestone for the systematics of this group.
Unseren ganz besonderen Dank wollen wir den 2 folgenden Personen aussprechen: Mrs. Patricia Gasperetti welche so freundlich war uns Zugang zu den Bildern ihres Ehegatten zu beschaffen, welcher im Jahre 2001 verstarb und dessen Beiträge immer noch Grundlage vieler Arbeiten zur arabischen Herpetofauna sind. Peter Brodmann jun., welcher uns behilflich war, und die Erlaubnis gab, Bilder seines Vaters, Peter Brodmann zu verwenden. Peter Brodmann sen. war ein großartiger Naturfreund und einer der besten Naturfotografen seiner Zeit. Er hinterließ uns das erste vollständige und unvergessliche Buch zu den Vipern Europas, das seiner Zeit weit voraus war, und das immer noch einen Meilenstein zur Systematik dieser Gruppe darstellt.

Lastly, we are indebted to Alexandre Teynié and Wolfgang Wüster for their constructive comments on the Introduction to this volume.
Zu guter Letzt wollen wir uns noch bei Alexandre Teynié and Wolfgang Wüster für ihre fachliche Kritik zu den taxonomischen Einleitungen bedanken.

Thanks to all of you.
Dank noch einmal an Euch alle.

RS09100-4 *Bungarus caeruleus* (SCHNEIDER, 1801)
Western Ghats, India
175 cm
photo: V. DEEPAK

RS09101-4 *Bungarus caeruleus* (SCHNEIDER, 1801)
Pakistan
175 cm
male
photo: G. VOGEL

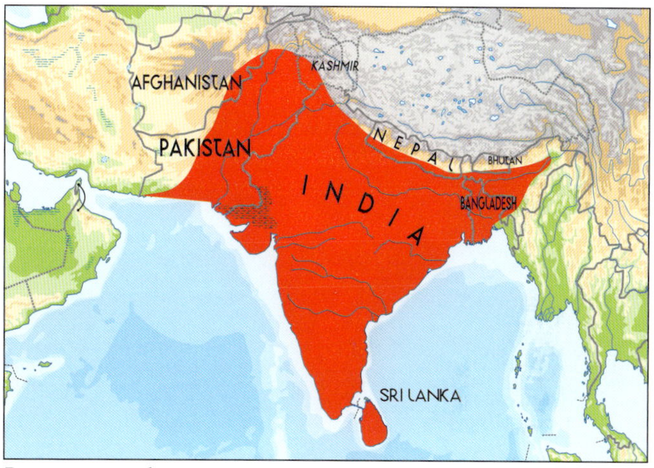

Bungarus caeruleus

RS09110-4 *Bungarus sindanus razai* KHAN, 1985
Dandot, Salt Range, Mianwali, Punjab Region,
Pakistan – 83 cm
photo: M.S. KHAN

Bungarus sindanus razai

RS09120-4 *Naja arabica* Scortecci, 1932
Yemen
240 cm
photo: A. Teynié

RS09121-4 *Naja arabica* Scortecci, 1932
Saudi Arabia
240 cm
photo: J. Gasperetti

Naja arabica

RS09123-4 *Naja arabica* SCORTECCI, 1932
map p. 31
Saudi Arabia
240 cm

photo: J. GASPERETTI

RS09130-4 *Naja oxiana* (EICHWALD, 1831)

163 cm

photo: L. TRUTNAU

RS09132-4 *Naja oxiana* (EICHWALD, 1831)
Uzbekistan
163 cm

photo: L. DARAFEYENKA

RS09131-4 *Naja oxiana* (EICHWALD, 1831)
Khorassan Province, Iran
163 cm

photo: H.G. KAMI

Naja oxiana

RS09300-4 *Walterinnesia aegyptia* LATASTE, 1887
Israel
135 cm
photo: A. BAR

RS09301-4 *Walterinnesia aegyptia* LATASTE, 1887
Israel
135 cm
photo: A. BAR

RS09302-4 *Walterinnesia aegyptia* LATASTE, 1887
Sinai Peninsula, Egypt
135 cm
photo: K. ROZINEK

RS09303 Habitat of *Atractaspis engaddensis, Walterinnesia aegyptia, Echis c. coloratus,* and *Pseudocerastes fieldii* photo: D. Jandzik
Wadi Feiran, Sinai Peninsula, Egypt

RS09304-4 *Walterinnesia aegyptia* Lataste, 1887
Jordan
135 cm
photo: D. Modrý

RS09305-4 *Walterinnesia aegyptia* Lataste, 1887
Jordan
135 cm
photo: D. Modrý

RS09306-4 *Walterinnesia aegyptia* Lataste, 1887
Sinai Peninsula, Egypt
135 cm
female
photo: D. Jandzik

RS09310-4 *Walterinnesia morgani* (Mocquard, 1905)

117 cm
photo: A. Leviton

Walterinnesia aegyptia (red)
Walterinnesia morgani (blue)

RS09311 Habitat of *Walterinnesia morgani*
Thumama, 100 km NW of Riyadh, Saudi Arabia
photo: T. Wilms

RS09312-2 *Walterinnesia morgani* (Mocquard, 1905), ZMGU.1150
Larestan, Fars Province, Iran
117 cm
juvenile
photo: H.G. Kami

RS09313-4 *Walterinnesia morgani* (Mocquard, 1905)
Thumama, 100 km NW of Riyadh, Saudi Arabia
117 cm
photo: T. Wilms

RS09400-4 *Atractaspis andersonii* Boulenger, 1905
Saudi Arabia
90 cm

photo: X. Eichaker

RS09401 Habitat of *Atractaspis andersonii*
Dhofar, Sultanate of Oman

photo: A Meyer

Atractaspis andersonii (blue)
Atractaspis engaddensis (red)

RS09410-4 *Atractaspis engaddensis* HAAS, 1950
map p. 27 Israel
65 cm

photo: A. BAR

RS09411-4 *Atractaspis engaddensis* HAAS, 1950
map p. 27 Negev, Israel
65 cm

photo: A. MEYER

RS009412 Habitat of *Atractaspis engaddensis*
Negev, Israel

photo: A. MEYER

RS09000-4 *Gloydius halys halys* (Pallas, 1776)
Balkhash Lake, Kazakhstan
75 cm
photo: J. de Smedt

RS09001-4 *Gloydius halys halys* (Pallas, 1776)
Balkhash Lake, Kazakhstan
75 cm
photo: J. de Smedt

RS09002-4 *Gloydius halys halys* (Pallas, 1776)
Western Kazakhstan
75 cm
male
photo: D. Jablonski

RS09003-4 *Gloydius halys halys* (Pallas, 1776)
Balkhash Lake, Kazakhstan
75 cm
photo: J. de Smedt

RS09004-4 *Gloydius halys halys* (Pallas, 1776)
Balkhash Lake, Kazakhstan
75 cm
photo: J. de Smedt

RS09005-4 *Gloydius halys halys* (Pallas, 1776)
Western Kazakhstan
75 cm
male
photo: D. Jablonski

Gloydius halys halys (red),
Gloydius halys boehmei (black)
Gloydius halys caucasicus (yellow)
Gloydius halys mogoi (blue)
Occurrence of both *Gloydius halys halys* and *Gloydius halys mogoi* (purple)

RS09010-4 *Gloydius halys boehmei* (Nilson, 1983), holotype
Andarab Valley, Baghlan Province, Afghanistan
44 cm
photo: A. Koch

RS09020-4 *Gloydius halys caucasicus* (NIKOLSKY, 1916) photo: B. TRAPP
Laar Valley near Teheran, Alborz (or Elburz) Mts., Iran
66 cm

RS09021-4 *Gloydius halys caucasicus* (NIKOLSKY, 1916)
Laar Valley near Teheran, Alborz (or Elburz) Mts.,
Iran – 66 cm
photo: B. TRAPP

RS09022-4 *Gloydius halys caucasicus* (NIKOLSKY, 1916)
Laar Valley near Teheran, Alborz (or Elburz) Mts.,
Iran – 66 cm
male
photo: D. JANDZIK

RS09023 Habitat of *Gloydius halys caucasicus*
Laar Valley near Teheran, Alborz (or Elburz) Mts., Iran
photo: B. TRAPP

RS09024-4 map p. 30 *Gloydius halys caucasicus* (NIKOLSKY, 1916)
Laar Valley near Teheran, Alborz (or Elburz) Mts., Iran – 66 cm
female
photo: D. JANDZIK

RS09025-4 map p. 30 *Gloydius halys caucasicus* (NIKOLSKY, 1916)
Sari City, Foresty Park, Mazandaran Province, Iran
66 cm
photo: H.G. KAMI

RS09030-4 map p. 30 *Gloydius halys mogoi* (BOUR, 1993)
Zabaikal'sky Kray, Russia
58 cm
photo: O. KORSUN

RS09031-4 map p. 30 *Gloydius halys mogoi* (BOUR, 1993)
Meadow, Khilogoson Village, Khiloksky District, Chitinskaya Oblast', Zabaikal'sky Kray, Russia – 76 cm
photo: O. KORSUN

RS09032-4 *Gloydius halys mogoi* (Bour, 1993)
map p. 30
Yedinenie, Chitinskaya Oblast', Zabaikal'sky Kray, Russia
76 cm
photo: O. Korsun

RS09033-4 *Gloydius halys mogoi* (Bour, 1993)
map p. 30
Pokrovka, Chitinskaya Oblast', Siberia, Russia
76 cm
photo: O. Korsun

RS09034 Habitat of *Gloydius halys mogoi*
Yapan Lake, valley of the Amur River, Mogochinsky District, Chitinskaya Oblast', Zabaikal'sky Kray, Eastern Siberia, Russia
photo: O. Korsun

RS09040-2 *Gloydius intermedius* (STRAUCH, 1868)

79 cm
juvenile
photo: D. JANDZIK

RS09041-4 *Gloydius intermedius* (STRAUCH, 1868)

79 cm
photo: D. HEGNER

RS09042-4 *Gloydius intermedius* (STRAUCH, 1868)
Korea
79 cm
photo: G. VOGEL

RS09043-4 *Gloydius intermedius* (STRAUCH, 1868)

79 cm
photo: W. WÜSTER

Gloydius intermedius

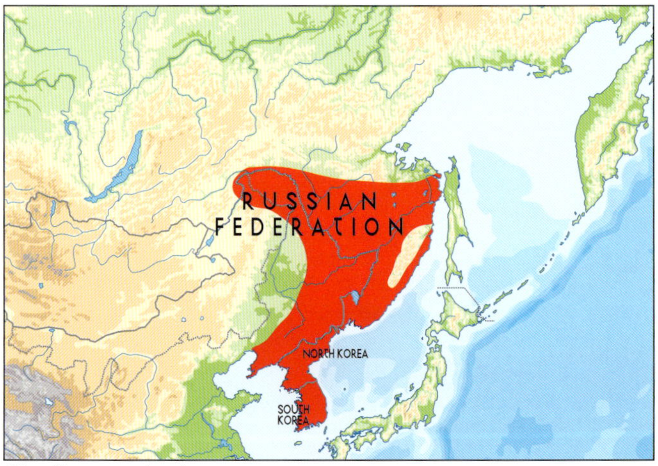

Gloydius ussuriensis

RS09050 Habitat of *Gloydius ussuriensis*
Yapan Lake, valley of the Amur River, Mogochinsky District, Chitinskaya Oblast',
Zabaikal'sky Kray, Eastern Siberia, Russia
photo: O. Korsun

RS09051-4 *Gloydius ussuriensis* (Emelianov, 1929)
Pokrovka, Zabaikal'sky Kray, Eastern Siberia, Russia
65 cm
photo: O. Korsun

RS09052-4 *Gloydius ussuriensis* (Emelianov, 1929)

65 cm
photo: L. Trutnau

RS09500-4 *Bitis arietans arietans* (Merrem, 1820)
near Taif, Saudi Arabia
191 cm

photo: N. Lutzmann

RS09501 Habitat of *Bitis arietans arietans*
near Taif, Saudi Arabia

photo: N. Lutzmann

RS09502-4 *Bitis arietans arietans* (Merrem, 1820)
near Taif, Saudi Arabia
191 cm

photo: N. Lutzmann

RS09503 Habitat of *Bitis arietans arietans*
Wadi Darbat, Salalah, Sultanate of Oman

photo: W. Wüster

RS09504-4 *Bitis arietans arietans* (Merrem, 1820)
near Taif, Saudi Arabia
191 cm

photo: T. Wilms

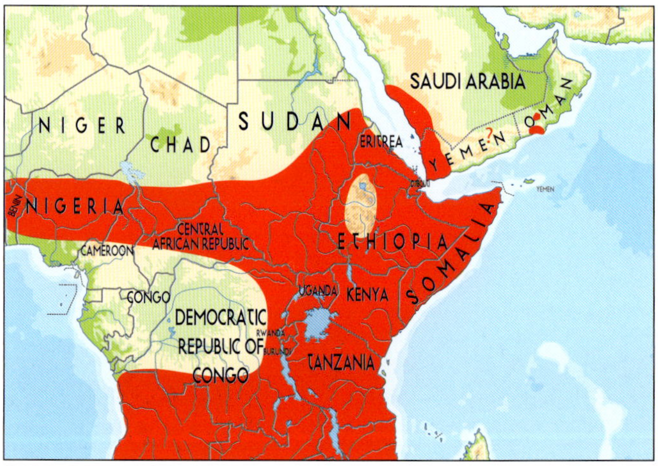

Bitis arietans arietans

RS09600-4 *Cerastes cerastes cerastes* (LINNAEUS, 1758)
Israel
89 cm
photo: A. BAR

RS09601-4 *Cerastes cerastes cerastes* (LINNAEUS, 1758)
Israel
89 cm
photo: A. BAR

RS09602-4 *Cerastes cerastes cerastes* (LINNAEUS, 1758)
Egypt
89 cm
photo: M. DOBIEY

RS09603-4 *Cerastes cerastes cerastes* (LINNAEUS, 1758)
Egypt
89 cm
male combat
photo: M. DOBIEY

RS09604-4 Tracks of *Cerastes* sp.
Israel
photo: A. BAR

RS09605-1 *Cerastes cerastes cerastes* (LINNAEUS, 1758)
89 cm
Hatching
photo: F. HULBERT

Cerastes cerastes cerastes (red)
Cerastes cerastes hoofieni (blue)

RS09606-4 *Cerastes cerastes cerastes* (LINNAEUS, 1758)
Egypt
89 cm
photo: D. JANDZIK

RS09610-4 *Cerastes cerastes hoofieni* WERNER, SIVAN, KUSHNIR & MOTRO, 1999 – Saudi Arabia
58 cm
photo: J. GASPERETTI

RS09620-4 *Cerastes gasperettii gasperettii* LEVITON & ANDERSON, 1967 – Jordan
86 cm

photo: D. MODRÝ

RS09621-4 *Cerastes gasperettii gasperettii* LEVITON & ANDERSON, 1967 – Jordan
86 cm

photo: D. MODRÝ

RS09622 Habitat of *Cerastes gasperettii gasperettii*
Al Wisad, Jordan

photo: D. MODRÝ

RS09623-4 *Cerastes gasperettii gasperettii* LEVITON & ANDERSON, 1967 – Sharjah, United Arab Emirates
86 cm

photo: K. KABISCH

RS09624-4 *Cerastes gasperettii gasperettii* LEVITON & ANDERSON, 1967 – Sharjah, United Arab Emirates
86 cm

photo: K. KABISCH

RS09625-4 *Cerastes gasperettii gasperettii* Leviton & Anderson, 1967 – Yemen
86 cm

photo: A. Teynié

RS09626-4 *Cerastes gasperettii gasperettii* Leviton & Anderson, 1967 – Jordan
86 cm

photo: D. Modrý

RS09630-4 *Cerastes gasperettii mendelssohni* Werner, Sivan, Kushnir & Motro, 1999 – Arava valley, next to the border with Jordan, Israel – 83 cm

photo: A. Bar

Cerastes gasperettii gasperettii (blue)
Cerastes gasperettii mendelssohni (violet)
Cerastes vipera (red)

RS09631-4 *Cerastes gasperettii mendelssohni* Werner, Sivan, Kushnir & Motro, 1999 – Arava valley, next to the border with Jordan, Israel – 83 cm

photo: A. Bar

RS09632-4 *Cerastes gasperettii mendelssohni* Werner, Sivan, Kushnir & Motro, 1999 – Arava valley, Israel
83 cm

photo: A. Meier

RS09633-4 *Cerastes gasperettii mendelssohni* WERNER, SIVAN, KUSHNIR & MOTRO, 1999
Arava valley, Israel
83 cm

photo: P. JANSSEN

RS09634-4 *Cerastes gasperettii mendelssohni* WERNER, SIVAN, KUSHNIR & MOTRO, 1999 – Arava valley, Israel
83 cm

photo: A. MEIER

RS09635 Habitat of *Cerastes gasperettii mendelssohni*
Israel

photo: P. JANSSEN

RS09640 Habitat of *Cerastes vipera*
Negev, Israel

photo: A. Meier

RS09641-4 *Cerastes vipera* (Linnaeus, 1758)
map p. 40 Egypt
48 cm
female

photo: D. Jandzik

RS09642-4 *Cerastes vipera* (Linnaeus, 1758)
map p. 40 Egypt
48 cm
female

photo: D. Jandzik

RS09643-4 *Cerastes vipera* (Linnaeus, 1758)
map p. 40 Negev, Israel
48 cm

photo: A. Meier

RS09644-4 *Cerastes vipera* (Linnaeus, 1758)
map p. 40 Egypt
48 cm
female

photo: D. Jandzik

RS09645-4 *Cerastes vipera* (LINNAEUS, 1758) photo: D. JABLONSKI
map p. 40 Sinai Peninsula, Egypt
48 cm

RS09646-4 *Cerastes vipera* (LINNAEUS, 1758)
map p. 40
48 cm
male
photo: F. FINKE

RS09647-4 *Cerastes vipera* (LINNAEUS, 1758)
map p. 40 Israel
48 cm
photo: A. BAR

RS09700-4 *Daboia palaestinae* (WERNER, 1938) photo: G. BABOCSAY
Jerusalem, photographed in a natural wadi, northern Jordan Valley, Israel
130 cm

RS09701-4 *Daboia palaestinae* (WERNER, 1938)
Israel
130 cm
photo: A. BAR

RS09702-4 *Daboia palaestinae* (WERNER, 1938)
Israel
130 cm
photo: A. BAR

RS09703 Habitat of *Daboia palaestinae* photo: D. Modrý
Ajlun, Jordan

Daboia palaestinae

RS09704-4 *Daboia palaestinae* (Werner, 1938)
Jordan
130 cm

photo: D. Modrý

RS09800-4 *Echis borkini* CHERLIN, 1990 photo: P. NEČAS
map p. 53 Al Khawkhah, Yemen

RS09801 Habitat of *Echis borkini*
Al Khawkhah, Yemen

photo: P. NEČAS

RS09802-4 *Echis borkini* CHERLIN, 1990
map p. 53 Al Khawkhah, Yemen

photo: P. NEČAS

RS09810-4 *Echis carinatus sochureki* Stemmler, 1969
Pakistan
61 cm
photo: D. Hegner

RS09811-4 *Echis carinatus sochureki* Stemmler, 1969
Pakistan
61 cm
photo: D. Hegner

RS09812-4 *Echis carinatus sochureki* Stemmler, 1969
near Angor, Sherabat, Southern Uzbekistan
61 cm
photo: T. Mazuch

RS09813-4 *Echis carinatus sochureki* STEMMLER, 1969
near Angor, Sherabat, Southern Uzbekistan
61 cm
photo: T. MAZUCH

RS09814-4 *Echis carinatus sochureki* STEMMLER, 1969
South-southeast of Nazwa or Nizwua, between Fathu and Shafa, Sultanate of Oman – 73 cm
photo: D. DONAIRE

RS09815-4 *Echis carinatus sochureki* STEMMLER, 1969
Sharjah, United Arab Emirates
73 cm
photo: W. WÜSTER

Echis carinatus sochureki (red)
The range of *Echis carinatus carinatus*, widespread in India, does not appear on this map

RS09816-4 *Echis carinatus sochureki* STEMMLER, 1969
Sharjah, United Arab Emirates
73 cm
photo: W. WÜSTER

RS09817-4 *Echis carinatus sochureki* STEMMLER, 1969
Southern Iran
73 cm
photo: V. GVOZDIK

RS09818-4 *Echis coloratus coloratus* (GÜNTHER, 1878)
Arava, Israel
90 cm
photo: A. MEIER

RS09820-4 *Echis coloratus coloratus* (GÜNTHER, 1878)
Israel
90 cm
photo: A. BAR

RS09821-4 *Echis coloratus coloratus* (Günther, 1878)
Israel
90 cm

photo: A. Bar

RS09822-4 *Echis coloratus coloratus* (Günther, 1878)
Egypt
90 cm
female

photo: D. Hegner

RS09824 Habitat of *Echis coloratus coloratus*
Jordan

photo: D. Modrý

RS09823-4 *Echis coloratus coloratus* (Günther, 1878)
vicinity of Thumrait, Dhofar, Sultanate of Oman
90 cm

photo: W. Wüster

RS09825-4 *Echis coloratus coloratus* (Günther, 1878)
Wadi Ramm, Jordan
90 cm

photo: D. Modrý

Echis coloratus coloratus (red)
Echis coloratus terraesanctae (blue)
Echis omanensis (purple)

RS09830-2 *Echis coloratus terraesanctae* BABOCSAY, 2003
near Jericho (Valley or River Jordan), Israel
59 cm
juvenile
photo: G. BABOCSAY

RS09831-4 *Echis coloratus* cf. *terraesanctae*
Negev, Israel
photo: A. MEIER

RS09832-4 *Echis coloratus* cf. *terraesanctae*
Negev, Israel
photo: A. MEIER

RS09833 Habitat of *Echis coloratus* cf. *terraesanctae*
Negev, Israel
photo: A. MEIER

RS09834-4 *Echis coloratus* cf. *terraesanctae*
Negev, Israel
photo: A. MEIER

RS09840-4 *Echis khosatzkii* Cherlin, 1990
Salalah, Dhofar, Sultanate of Oman
40 cm

photo: W. Wüster

RS09841-4 *Echis khosatzkii* Cherlin, 1990
Salalah, Dhofar, Sultanate of Oman
40 cm

photo: W. Wüster

RS09842-4 *Echis khosatzkii* Cherlin, 1990
15 km NW of Qairun Hairitti, Dhofar, Sultanate of Oman – 40 cm

photo: D. Donaire

RS09843 Habitat of *Echis khosatzkii*
Dhofar, Sultanate of Oman

photo: A. Meyer

RS09844-4 *Echis khosatzkii* Cherlin, 1990
Dhofar, Sultanate of Oman
40 cm

photo: A. Meyer

Echis khosatzkii (blue)
Echis borkini (red)

RS09850-4 *Echis omanensis* Babocsay, 2004
map p. 51 Al Jabal al Akhdar, Sultanate of Oman
61 cm

photo: A. Meyer

RS09851-4 *Echis omanensis* Babocsay, 2004
map p. 51 Al Jabal al Akhdar, Sultanate of Oman
61 cm

photo: A. Meyer

RS09853 Habitat of *Echis omanensis*
Hatta, United Arab Emirates

photo: W. Wüster

RS09852-4 *Echis omanensis* Babocsay, 2004
map p. 51 Bani Habib, Sultanate of Oman
61 cm

photo: D. Donaire

RS09854-4 *Echis omanensis* Babocsay, 2004
map p. 51 Hatta, United Arab Emirates
61 cm

photo: W. Wüster

Eristicophis macmahonii

RS09900-4 *Eristicophis macmahonii* ALCOCK & FINN, 1897
Pakistan
72 cm
male
photo: D. JANDZIK

RS09901-4 *Eristicophis macmahonii* ALCOCK & FINN, 1897
Pakistan
72 cm
female
photo: D. JANDZIK

RS10000 Habitat of *Macrovipera lebetina lebetina*
Troodos Mts., Cyprus

photo: K. KABISCH

RS10001-4 *Macrovipera lebetina lebetina* (LINNAEUS, 1758)
Troodos Mts., Cyprus
153 cm

photo: K. KABISCH

RS10002-4 *Macrovipera lebetina lebetina* (LINNAEUS, 1758)
Cyprus
153 cm

photo: F. BAIER

RS10003-4 *Macrovipera lebetina lebetina* (LINNAEUS, 1758)
Kaleburnu, Northern Cyprus
153 cm

photo: H. SIGG

RS10004-4 *Macrovipera lebetina lebetina* (LINNAEUS, 1758)
Paphos, Cyprus
153 cm
female

photo: A. MEYER

RS10005-4 *Macrovipera lebetina lebetina* (LINNAEUS, 1758)
Paphos, Cyprus
153 cm
female
photo: A. MEYER

Macrovipera lebetina lebetina (yellow)
Macrovipera lebetina cernovi (blue)
Macrovipera lebetina euphratica (purple)
Macrovipera lebetina obtusa (red)
Macrovipera lebetina turanica (pink)
Macrovipera schweizeri (black arrow)

RS10010-4 *Macrovipera lebetina cernovi* CHIKIN & SZCZERBAK, 1992 – Sarigol National Park, Northern Khorasan Province, Iran
125 cm
photo: B. JAFARI & H.G. KAMI

RS10011 Habitat of *Macrovipera lebetina cernovi*
Sarigol National Park, Northern Khorasan Province, Iran
photo: B. Jafari & H.G. Kami

RS10012-4 *Macrovipera lebetina cernovi* Chikin & Szczerbak, 1992 – Sarigol National Park, Northern Khorasan Province, Iran
map p. 57
125 cm
photo: B. Jafari & H.G. Kami

RS10013-4 *Macrovipera lebetina cernovi* Chikin & Szczerbak, 1992 – Sarigol National Park, Northern Khorasan Province, Iran
map p. 57
125 cm
photo: B. Jafari & H.G. Kami

RS10020-4 *Macrovipera lebetina euphratica* (MARTIN, 1838)
map p. 57 Adana Province, Turkey
ca. 150 cm
photo: P. GENIEZ & A. TEYNIÉ

RS10021-4 *Macrovipera lebetina euphratica* (MARTIN, 1838)
map p. 57 Antakya Province, Turkey
ca. 150 cm
photo: A. TEYNIÉ

RS10022-4 *Macrovipera lebetina euphratica* (MARTIN, 1838)
map p. 57 Antakya Province, Turkey
ca. 150 cm
photo: P. GENIEZ & A. TEYNIÉ

RS10023-4 *Macrovipera lebetina euphratica* (MARTIN, 1838)
map p. 57 Jordan
ca 150 cm
photo: D. MODRÝ

RS10030 Habitat of *Macrovipera lebetina obtusa*
Qobustan, Azerbaijan
photo: V. GVOZDIK

RS10031-2 *Macrovipera lebetina obtusa* (DWIGUBSKY, 1832)
map p. 57 Qobustan, Azerbaijan
214 cm
juvenile
photo: V. GVOZDIK

RS10032-2 *Macrovipera lebetina obtusa* (DWIGUBSKY, 1832) photo: V. GVOZDIK
map p. 57 Qobustan, Azerbaijan
214 cm
juvenile

RS10033 Habitat of *Marcovipera lebetina obtusa*
Quasvin, Iran
photo: B. TRAPP

RS10034-4 *Macrovipera lebetina obtusa* (DWIGUBSKY, 1832)
map p. 57 near Quasvin, Iran
214 cm
photo: B. TRAPP

RS10035-4 *Macrovipera lebetina obtusa* (Dwigubsky, 1832)
map p. 57 near Quasvin, Iran
214 cm
photo: B. Trapp

RS10036-2 *Macrovipera lebetina obtusa* (Dwigubsky, 1832)
map p. 57 Artwin Province, Turkey
214 cm
juvenile
photo: P. Geniez & A. Teynié

RS10037-4 *Macrovipera lebetina obtusa* (Dwigubsky, 1832)
map p. 57 Kars Province, Turkey
214 cm
female
photo: P. Geniez & A. Teynié

RS10038-4 *Macrovipera lebetina obtusa* (Dwigubsky, 1832)
map p. 57 between Erzurum and Diyarbakir, Eastern Turkey
214 cm
photo: J. de Smedt

RS10039 Habitat of *Macrovipera lebetina obtusa*
Kars Province, Turkey
photo: P. Geniez & A. Teynié

RS10040-4 *Macrovipera lebetina turanica* (CHERNOV *in* TERENT'EV & CHERNOV, 1940) – Nura Tau, Uzbekistan
map p. 57
158 cm
photo: L. WESTRIN

RS10041-4 *Macrovipera lebetina turanica* (CHERNOV *in* TERENT'EV & CHERNOV, 1940) – Nura Tau, Uzbekistan
map p. 57
158 cm
photo: D. JABLONSKI

RS10042-4 *Macrovipera lebetina turanica* (CHERNOV *in* TERENT'EV & CHERNOV, 1940) – Nura Tau, Uzbekistan
map p. 57
158 cm
photo: T. MASUCH

RS10043-4 *Macrovipera lebetina turanica* (CHERNOV *in* TERENT'EV & CHERNOV, 1940)
map p. 57
photo: L. TRUTNAU

RS10050-2 *Macrovipera schweizeri* (WERNER, 1935)
map p. 57
Milos, Cyclades Islands, Greece
107 cm
juvenile
photo: M. DOBIEY

RS10051-4 *Macrovipera schweizeri* (WERNER, 1935)
map p. 57
Milos, Cyclades Islands, Greece
107 cm
male
photo: M. DOBIEY

RS10052-4 *Macrovipera schweizeri* (WERNER, 1935)
map p. 57 Milos, Cyclades Islands, Greece
107 cm
photo: A. MEYER

RS10053-4 *Macrovipera schweizeri* (WERNER, 1935)
map p. 57 Milos, Cyclades Islands, Greece
107 cm
photo: A. MEYER

RS10054-4 *Macrovipera schweizeri* (WERNER, 1935)
map p. 57 Milos, Cyclades Islands, Greece
107 cm
photo: H. BILLING

RS10055 Habitat of *Macrovipera schweizeri*
Milos, Cyclades Islands, Greece
photo: H. BILLING

RS10056-4 *Macrovipera schweizeri* (WERNER, 1935)
map p. 57
Milos, Cyclades Islands, Greece
107 cm
photo: H. BILLING

RS10057-4 *Macrovipera schweizeri* (WERNER, 1935)
map p. 57
Milos, Cyclades Islands, Greece
107 cm
photo: E. EVAN

RS10058-4 *Macrovipera schweizeri* (WERNER, 1935)
map p. 57 Milos, Cyclades Islands, Greece
107 cm
photo: B. TRAPP

RS10059-4 *Macrovipera schweizeri* (WERNER, 1935)
map p. 57 Milos, Cyclades Islands, Greece
107 cm
photo: J. VAN DER VOORT

RS10060-4 *Macrovipera schweizeri* (WERNER, 1935)
map p. 57 Milos, Cyclades Islands, Greece
107 cm
photo: E. EVAN

RS10061-4 *Macrovipera schweizeri* (WERNER, 1935), red morph
map p. 57 Milos, Cyclades Islands, Greece
107 cm
photo: B. TRAPP

RS10062-4 *Macrovipera schweizeri* (WERNER, 1935), red morph
map p. 57 Milos, Cyclades Islands, Greece
107 cm
photo: O. KLINGNER

RS10100-4 *Montivipera albicornuta* (Nilson & Andrén, 1985) photo: B. Trapp
map p. 74 Zanjan, Alborz (or Elburz) Mts., Iran
66 cm

RS10101 Habitat of *Montivipera albicornuta*, type locality
Zanjan, Alborz (or Elburz) Mts., Iran
photo: B. Trapp

RS10102-4 *Montivipera albicornuta* (Nilson & Andrén, 1985)
map p. 74 Zanjan, Alborz (or Elburz) Mts., Iran
66 cm
photo: B. Trapp

RS10103-4 *Montivipera albicornuta* (NILSON & ANDRÉN, 1985)
map p. 74 Zanjan, Alborz (or Elburz) Mts., Iran
66 cm

photo: B. TRAPP

RS10110-4 *Montivipera albizona* (NILSON, ANDRÉN & FLÄRDH, 1990) – Sivas Province, Turkey
map p. 74
78 cm

photo: P. GENIEZ & A. TEYNIÉ

RS10111-4 *Montivipera albizona* (NILSON, ANDRÉN & FLÄRDH, 1990)
map p. 74 Sivas Province, Turkey
78 cm
male

photo: P. GENIEZ & A. TEYNIÉ

RS10112 Habitat of *Montivipera albizona*
Tecer Daglari, Central Turkey

photo: D. Jandzik

RS10113-4 map p. 74
Montivipera albizona (Nilson, Andrén & Flärdh, 1990) – Tecer Daglari, Central Turkey
78 cm
male
photo: D. Jandzik

RS10114-4 map p. 74
Montivipera albizona (Nilson, Andrén & Flärdh, 1990) – Tecer Daglari, Central Turkey
78 cm
female
photo: D. Jandzik

RS10115-4 map p. 74
Montivipera albizona (Nilson, Andrén & Flärdh, 1990) – Sivas Province, Turkey
78 cm
female
photo: P. Geniez & A. Teynié

RS10116-4 map p. 74
Montivipera albizona (Nilson, Andrén & Flärdh, 1990) – Sivas Province, Turkey
78 cm
male
photo: P. Geniez & A. Teynié

RS10120-4 *Montivipera bornmuelleri* (Werner, 1898)
map p. 74
Israel
74 cm

photo: A. Bar

RS10121-4 *Montivipera bornmuelleri* (Werner, 1898)
map p. 74
Israel
74 cm

photo: A. Bar

RS10122 Habitat of *Montivipera bornmuelleri*
Mt. Hermon, Israel

photo: A. Bar

RS10123-4 *Montivipera bornmuelleri* (WERNER, 1898)
map p. 74
74 cm
photo: L. TRUTNAU

RS10124-4 *Montivipera bornmuelleri* (WERNER, 1898)
map p. 74 Mt. Hermon, Israel
74 cm
male
photo: G. BABOCSAY

RS10130-4 *Montivipera bulgardaghica* (NILSON & ANDRÉN, 1985)
map p. 74 Adana Province, Turkey
78 cm
photo: A. TEYNIÉ

RS10131-4 *Montivipera bulgardaghica* (NILSON & ANDRÉN, 1985)
map p. 74 Adana Province, Turkey
78 cm
photo: A. TEYNIÉ

RS10132-4 *Montivipera bulgardaghica* (NILSON & ANDRÉN, 1985)
map p. 74 Adana Province, Turkey
78 cm
photo: A. TEYNIÉ

RS10133 Habitat of *Montivipera bulgardaghica*
Turkey
photo: F. SIEGENTHALER

RS10140-4 *Montivipera latifii* (MERTENS, DAREVSKY & KLEMMER, 1967)
map p. 74
79 cm
female
photo: M. KARLSSON

RS10141-4 *Montivipera latifii* (MERTENS, DAREVSKY & KLEMMER, 1967)
map p. 74
79 cm
female
photo: M. KARLSSON

RS10142-4 *Montivipera latifii* (MERTENS, DAREVSKY & KLEMMER, 1967) – Laar Valley, Alborz (or Elburz) Mts., Teheran Province, Iran – 79 cm
map p. 74
photo: M. RAJABIZADEH

RS10143-4 *Montivipera latifii* (MERTENS, DAREVSKY & KLEMMER, 1967) – Laar Valley, Alborz (or Elburz) Mts., Teheran Province, Iran – 79 cm
map p. 74
photo: M. RAJABIZADEH

RS10145 Habitat of *Montivipera latifii*
Laar Valley, Alborz (or Elburz) Mts., Teheran Province, Iran
photo: M. RAJABIZADEH

RS10146-4 *Montivipera latifii* (MERTENS, DAREVSKY & KLEMMER, 1967) – Iran
map p. 74
79 cm
photo: J. TASHJIAN

RS10147-4 *Montivipera latifii* (MERTENS, DAREVSKY & KLEMMER, 1967) – Iran
map p. 74
79 cm
photo: J. TASHJIAN

RS10148-4 *Montivipera latifii* (MERTENS, DAREVSKY & KLEMMER, 1967) – Iran
map p. 74
79 cm
photo: J. TASHJIAN

RS10149-4 *Montivipera latifii* (MERTENS, DAREVSKY & KLEMMER, 1967) – Laar Valley, Alborz (or Elburz) Mts., Teheran Province, Iran – 79 cm
map p. 74
female
photo: P. BRODMANN

RS10150-4 *Montivipera latifii* (MERTENS, DAREVSKY & KLEMMER, 1967) – Laar Valley, Alborz (or Elburz) Mts., Teheran Province, Iran – 79 cm
map p. 74
female
photo: P. BRODMANN

RS10160-4 *Montivipera raddei raddei* (BOETTGER, 1890)
map p. 74
Digor, Turkey
99 cm
gravid female
photo: D. JANDZIK

RS10161-2 *Montivipera raddei raddei* (BOETTGER, 1890)
map p. 74
Digor, Turkey
99 cm
juveniles
photo: D. JANDZIK

RS10162-4 *Montivipera raddei raddei* (Boettger, 1890) photo: D. Bohle
map p. 74 Armenia
99 cm

RS10163-4 *Montivipera raddei raddei* (Boettger, 1890)
map p. 74
99 cm
female photo: D. Jandzik

RS10164-4 *Montivipera raddei raddei* (Boettger, 1890)
map p. 74
99 cm
photo: L. Trutnau

RS10174-4 *Montivipera raddei raddei* (Boettger, 1890) photo: D. Bohle
Armenia
99 cm

Montivipera albicornuta (black), *Montivipera albizona* (brown), *Montivipera bornmuelleri* (orange), *Montivipera bulgardaghica* (blue), *Montivipera latifii* (green), *Montivipera raddei raddei* (olive green), *Montivipera raddei kurdistanica* (pink), *Montivipera wagneri* (yellow), *Montivipera xanthina* (red)

RS10166-4 *Montivipera raddei raddei* (Boettger, 1890)
Mt. Ararat, Turkey
99 cm
male
photo: A. Teynié

RS10167-4 *Montivipera raddei raddei* (Boettger, 1890)
Mt. Ararat, Turkey
99 cm
male
photo: D. Jablonski

RS10168 Habitat of *Montivipera raddei raddei*
Mt. Ararat, Turkey
photo: A. Teynié

RS10169-2 *Montivipera raddei raddei* (Boettger, 1890)
Mt. Ararat, Turkey
99 cm
juvenile
photo: P. Geniez & A. Teynié

RS10170-4 *Montivipera raddei raddei* (BOETTGER, 1890)
map p. 74
99 cm
female
photo: H. WILHELM

RS10171-2 *Montivipera raddei raddei* (BOETTGER, 1890)
map p. 74
99 cm
juveniles
photo: D. JANDZIK

RS10172-4 *Montivipera raddei raddei* (BOETTGER, 1890)
map p. 74
Iran
photo: K. RAJABIZADEH

RS10173-4 *Montivipera raddei raddei* (BOETTGER, 1890)
map p. 74
Armenia
99 cm
female
photo: P. BRODMANN

RS10180-2 *Montivipera raddei kurdistanica* (NILSON & ANDRÉN, 1986) – Hakkari Province, Turkey
map p. 74
89 cm
juvenile
photo: G. VOGEL

RS10181-4 *Montivipera raddei kurdistanica* (NILSON & ANDRÉN, 1986)
map p. 74
89 cm
photo: L. TRUTNAU

RS10190-4 *Montivipera wagneri* (Nilson & Andrén, 1984) photo: D. Jandzik
map p. 74 Karakurt, Turkey
95 cm
male

RS10191-4 *Montivipera wagneri* (Nilson & Andrén, 1984)
map p. 74 Karakurt, Turkey
95 cm
male photo: D. Jandzik

RS10192-4 *Montivipera wagneri* (Nilson & Andrén, 1984)
map p. 74 Karakurt, Turkey
95 cm
male photo: D. Jandzik

RS10193 *Montivipera wagneri* (Nilson & Andrén, 1984)
presumable type locality of the species
Ars Valley, Kars Province, Turkey
photo: B. Trapp

RS10194-4 *Montivipera wagneri* (Nilson & Andrén, 1984)
map p. 74 Karakurt, Turkey
95 cm
male
photo: A. Teynié

RS10195-4 *Montivipera wagneri* (Nilson & Andrén, 1984)
map p. 74 Karakurt, Turkey
95 cm
female
photo: P. Geniez & A. Teynié

RS10196-4 *Montivipera wagneri* (Nilson & Andrén, 1984)
map p. 74 North of Lake Van, Turkey
95 cm
female
photo: P. Brodmann

RS10197-4 *Montivipera wagneri* (Nilson & Andrén, 1984)
map p. 74 Karakurt, Turkey
95 cm
female
photo: A. Teynié

RS10200 Habitat of *Montivipera xanthina*
Ortaklar, Western Turkey
photo: K. Mebert

RS10201-4 *Montivipera xanthina* (Gray, 1849)
map p. 74
Ortaklar, Western Turkey
116 cm
female
photo: K. Mebert

XI → D

RS10202-4 *Montivipera xanthina* (Gray, 1849)
map p. 74
Ortaklar, Western Turkey
116 cm
female
photo: K. Mebert

XI → D

RS10203-4 *Montivipera xanthina* (GRAY, 1849)
map p. 74 Tekir Dag, Turkey (European part)
116 cm
photo: F. SIEGENTHALER

RS10204-4 *Montivipera xanthina* (GRAY, 1849)
map p. 74 Olympos near Antalya, Turkey
116 cm
photo: F. SIEGENTHALER

RS10205-4 *Montivipera xanthina* (GRAY, 1849)
map p. 74 Tekir Dag, Turkey (European part)
116 cm
photo: F. SIEGENTHALER

RS10206-4 *Montivipera xanthina* (GRAY, 1849)
map p. 74 Turkey
116 cm
male
photo: D. JANDZIK

RS10207-4 *Montivipera xanthina* (GRAY, 1849)
map p. 74 Aydin Province, Turkey
116 cm
photo: A. AVCI

RS10208-4 *Montivipera xanthina* (Gray, 1849)
map p. 74 Kayseri Province, Turkey
116 cm
male
photo: P. Geniez & A. Teynié

RS10209-4 *Montivipera xanthina* (Gray, 1849)
map p. 74 Thráki, Greece
116 cm
female
photo: A. Meyer

RS10210-4 *Montivipera xanthina* (Gray, 1849)
map p. 74 Thráki, Greece
116 cm
male
photo: A. Meyer

RS10211 Habitat of *Montivipera xanthina*
Samos Island, Greece

photo: A. Meyer

RS10212-4 *Montivipera xanthina* (Gray, 1849)
map p. 74 Samos Island, Greece
116 cm
female

photo: A. Meyer

RS10213-4 *Montivipera xanthina* (Gray, 1849)
map p. 74 Lipsoì, Greece
116 cm
female

photo: A. Meyer

RS10214-4 *Montivipera xanthina* (Gray, 1849)
map p. 74 Samos Island, Greece
116 cm

photo: A. Meyer

RS10215-4 *Montivipera xanthina* (Gray, 1849)
map p. 74 surroundings of Alexandropolis, Evros, Greece
116 cm

photo: J. van der Voort

RS10216-4 *Montivipera xanthina* (Gray, 1849)
map p. 74 surroundings of Alexandropolis, Evros, Greece
116 cm

photo: J. van der Voort

RS10300-4 *Pseudocerastes fieldii* SCHMIDT, 1930
Negev, Israel
89 cm

photo: A. MEYER

RS10301 Habitat of *Pseudocerastes fieldii*
Palmyra, Syria

photo: D. JANDZIK

RS10302-4 *Pseudocerastes fieldii* SCHMIDT, 1930
Negev, Israel
89 cm

photo: A. MEYER

RS10303-4 *Pseudocerastes* cf. *fieldii*
Lava desert, Syria
? cm
female
photo: D. Jandzik

Pseudocerastes persicus (red)
Pseudocerastes fieldii (blue)
Pseudocerastes urarachnoides (purple)

RS10304-4 *Pseudocerastes* cf. *fieldii*
Lava desert, Syria
? cm
female
photo: D. Jandzik

RS10310-2 *Pseudocerastes persicus* (Duméril, Bibron & Duméril, 1854) – Miandasht wildlife refuge, Northern Khorasan Province, Iran – 108 cm juvenile
photo: B. Jafari & H.G. Kami

RS10311 Habitat of *Pseudocerastes persicus*
Miandasht wildlife refuge, Northern Khorasan Province, Iran
photo: B. Jafari & H.G. Kami

RS10312-4 *Pseudocerastes persicus* (Duméril, Bibron & Duméril, 1854) – Pakistan
108 cm
photo: D. Jandzik

RS10320 Habitat of *Pseudocerastes urarachnoides*
Area of Kooleg, south of Chakar, Bina, Iran
photo: B. Fathinia

RS10321-4 *Pseudocerastes urarachnoides* Bostanchi, Anderson, Kami & Papenfuss, 2006 – Area of Kooleg, south of Chakar, Bina, Iran – 53 cm
photo: B. Fathinia

RS10322-4 *Pseudocerastes urarachnoides* Bostanchi, Anderson, Kami & Papenfuss, 2006 – Area of Kooleg, south of Chakar, Bina, Iran – 53 cm
photo: B. Fathinia

RS10400-4 *Vipera ammodytes ammodytes* (LINNAEUS, 1758)
map p. 92 Novomesto, Jogorie, Slovenia
110 cm
male
photo: T. BÖHM

RS10401-4 *Vipera ammodytes ammodytes* (LINNAEUS, 1758)
map p. 92 Slovenia
110 cm
female
photo: A. MEYER

RS10402-4 *Vipera ammodytes ammodytes* (LINNAEUS, 1758)
map p. 92 Carinthia, Austria
110 cm
female
photo: C. RIEGLER

RS10403-4 *Vipera ammodytes ammodytes* (LINNAEUS, 1758)
map p. 92 Carinthia, Austria
110 cm
male
photo: C. RIEGLER

RS10404-4 *Vipera ammodytes ammodytes* (LINNAEUS, 1758)
map p. 92 Carinthia, Austria
110 cm
male
photo: A. MEYER

RS10405-4 *Vipera ammodytes ammodytes* (LINNAEUS, 1758)
map p. 92 North Italy
110 cm
male
photo: D. JANDZIK

RS10406-4 *Vipera ammodytes ammodytes* (Linnaeus, 1758) photo: K. Mebert
map p. 92 Ada Island, Montenegro
110 cm

RS10407-4 *Vipera ammodytes ammodytes* (Linnaeus, 1758)
map p. 92 Ada Island, Montenegro
110 cm
male photo: T. Ott

RS10408-4 *Vipera ammodytes ammodytes* (Linnaeus, 1758)
map p. 92 Skadarsko Lake, Vranjina Island, Montenegro
110 cm
female photo: M. Schweiger

RS10409-4 *Vipera ammodytes ammodytes* (LINNAEUS, 1758)
map p. 92 Montenegro
110 cm
female
photo: T. OTT

RS10410-4 *Vipera ammodytes ammodytes* (LINNAEUS, 1758)
map p. 92 Montenegro
110 cm
female
photo: T. OTT

RS10411-4 *Vipera ammodytes ammodytes* (LINNAEUS, 1758)
map p. 92 Montenegro
110 cm
male
photo: T. OTT

RS10412-4 *Vipera ammodytes ammodytes* (LINNAEUS, 1758)
map p. 92 Montenegro
110 cm
male
photo: T. OTT

RS10413-4 *Vipera ammodytes ammodytes* (LINNAEUS, 1758)
map p. 92 Montenegro
110 cm
male
photo: T. OTT

RS10414-4 *Vipera ammodytes ammodytes* (LINNAEUS, 1758)
map p. 92 Ulcinj, Montenegro
110 cm
photo: D. BOHLE

RS10415-4 *Vipera ammodytes ammodytes* (LINNAEUS, 1758)
map p. 92 Montenegro
110 cm
male
photo: T. OTT

RS10416 Habitat of *Vipera ammodytes ammodytes*
Ulcinj, Montenegro

photo: D. Bohle

RS10417-4 *Vipera ammodytes ammodytes* (Linnaeus, 1758)
map p. 92 Ulcinj, Montenegro
110 cm

photo: D. Bohle

RS10418-4 *Vipera ammodytes ammodytes* (Linnaeus, 1758)
map p. 92 Ulcinj, Montenegro
110 cm

photo: D. Bohle

RS10419-4 *Vipera ammodytes ammodytes* (Linnaeus, 1758)
map p. 92 Ulcinj, Montenegro
110 cm

photo: D. Bohle

RS10420-4 *Vipera ammodytes ammodytes* (Linnaeus, 1758)
map p. 92 Ulcinj, Montenegro
110 cm

photo: D. Bohle

RS10421-4 *Vipera ammodytes ammodytes* (Linnaeus, 1758)
map p. 92 Ulcinj, Montenegro
110 cm

photo: D. Bohle

RS10422-4 *Vipera ammodytes ammodytes* (LINNAEUS, 1758)
map p. 92
Skutari Lake, Montenegro
110 cm
female
photo: S. DUMMERMUTH

RS10423 Habitat of *Vipera ammodytes ammodytes*
Skutari Lake, Montenegro
photo: O. KLINGNER

RS10424-2 *Vipera ammodytes ammodytes* (LINNAEUS, 1758)
map p. 92
Skutari Lake, Montenegro
110 cm
juvenile male
photo: S. DUMMERMUTH

RS10425-4 *Vipera ammodytes ammodytes* (LINNAEUS, 1758)
map p. 92
Skutari Lake, Montenegro
110 cm
male
photo: S. DUMMERMUTH

RS10426-4 *Vipera ammodytes ammodytes* (LINNAEUS, 1758)
map p. 92
Skutari Lake, Montenegro
110 cm
male
photo: S. DUMMERMUTH

RS10427-4 *Vipera ammodytes ammodytes* (LINNAEUS, 1758)
map p. 92
Skutari Lake, Montenegro
110 cm
male
photo: S. DUMMERMUTH

RS10428 Habitat of *Vipera ammodytes ammodytes* (form *gregorwallneri*) – Carinthia, Austria

photo: J. Hill

RS10429-4 *Vipera ammodytes ammodytes* (Linnaeus, 1758) (form *gregorwallneri*) – Carinthia, Austria
110 cm

photo: J. Hill

RS10430-4 *Vipera ammodytes ammodytes* (Linnaeus, 1758) (form *gregorwallneri*) – Carinthia, Austria
110 cm
female

photo: P. Janssen

RS10431-4 *Vipera ammodytes ammodytes* (Linnaeus, 1758) (form *gregorwallneri*) – Carinthia, Austria
110 cm
male

photo: D. Jablonski

Vipera ammodytes ammodytes (red)
Vipera ammodytes meridionalis (blue)
Vipera ammodytes montandoni (purple)
Vipera ammodytes transcaucasiana (black)

RS10432 Habitat of *Vipera ammodytes ammodytes* (form *ruffoi*) – Bolzano, Italy

photo: O. Klingner

RS10433-4 *Vipera ammodytes ammodytes* (Linnaeus, 1758) (form *ruffoi*) – Bolzano, Italy
110 cm
male

photo: D. Bohle

RS10434-4 *Vipera ammodytes ammodytes* (Linnaeus, 1758) (form *ruffoi*) – Bolzano, Italy
110 cm

photo: D. Bohle

RS10435-4 *Vipera ammodytes ammodytes* (Linnaeus, 1758) (form *ruffoi*) – Bolzano, Italy
110 cm
male

photo: A. Pieh

RS10436-4 *Vipera ammodytes ammodytes* (Linnaeus, 1758) (form *ruffoi*) – Bolzano, Italy
110 cm
female

photo: T. Ott

RS10437-4 *Vipera ammodytes ammodytes* (Linnaeus, 1758) (form *ruffoi*) – Bolzano, Italy
110 cm
female

photo: T. Ott

RS10480 Habitat of *Vipera ammodytes meridionalis*
Shelegur Lake, Southeastern Albania
photo: D. Jablonski

RS10481-4 *Vipera ammodytes meridionalis* Boulenger, 1903
map p. 92
Albania
65 cm
gravid female
photo: D. Jablonski

RS10482-2 *Vipera ammodytes meridionalis* Boulenger, 1903
map p. 92
Skutari Lake, Albania
65 cm
juvenile male
photo: T. Mazuch

RS10483 Habitat of *Vipera ammodytes meridionalis*
Southwest Bulgaria
photo: D. Jandzik

RS10484-4 *Vipera ammodytes meridionalis* Boulenger, 1903
map p. 92
Southwest Bulgaria
65 cm
gravid female
photo: D. Jandzik

RS10485-4 *Vipera ammodytes meridionalis* Boulenger, 1903
map p. 92
Ropotamo, South Bulgaria
65 cm
photo: J. van der Voort

RS10486-4 *Vipera ammodytes meridionalis* BOULENGER, 1903
map p. 92 Makedonia, Greece
65 cm
male
photo: A. MEYER

RS10487-4 *Vipera ammodytes meridionalis* BOULENGER, 1903
map p. 92 Ípeiros, Greece
65 cm
male
photo: A. MEYER

RS10488-4 *Vipera ammodytes meridionalis* BOULENGER, 1903
map p. 92 Ípeiros, Greece
65 cm
male
photo: A. MEYER

RS10489-4 *Vipera ammodytes meridionalis* BOULENGER, 1903
map p. 92 Kardamyli, Messinia, Peloponnese, Greece
65 cm
photo: J. VAN DER VOORT

RS10490 Habitat of *Vipera ammmodytes meridionalis*
Morfi Marsh, Greece

photo: S. Dummermuth

RS10491-4 *Vipera ammmodytes meridionalis* Boulenger, 1903
map p. 92 Morfi Marsh, Greece
65 cm
male

photo: S. Dummermuth

RS10492-4 *Vipera ammmodytes meridionalis* Boulenger, 1903
map p. 92 Morfi Marsh, Greece
65 cm
male

photo: S. Dummermuth

RS10493-4 *Vipera ammmodytes meridionalis* Boulenger, 1903
map p. 92 Morfi Marsh, Greece
65 cm
female

photo: S. Dummermuth

RS10494-4 *Vipera ammodytes meridionalis* Boulenger, 1903
map p. 92 Makedonia, Greece
65 cm
female

photo: A. Meyer

RS10495-4 map p. 92 *Vipera ammodytes meridionalis* BOULENGER, 1903
Makedonia, Greece
65 cm
female
photo: A. MEYER

RS10496-4 map p. 92 *Vipera ammodytes meridionalis* BOULENGER, 1903
Ípeiros, Greece
65 cm
photo: A. MEYER

RS10497-4 map p. 92 *Vipera ammodytes meridionalis* BOULENGER, 1903
vicinity of Athens, Attica, Greece
65 cm
photo: J. VAN DER VOORT

RS10498-4 map p. 92 *Vipera ammodytes meridionalis* BOULENGER, 1903
Thráki, Greece
65 cm
female
photo: A. MEYER

RS10499-4 map p. 92 *Vipera ammodytes meridionalis* BOULENGER, 1903
Kardamyli, Messinia, Peloponnese, Greece
65 cm
photo: J. VAN DER VOORT

RS10500-4 map p. 92 *Vipera ammodytes meridionalis* BOULENGER, 1903
Peloponnese, Greece
65 cm
photo: J. HILL

RS10501-4 *Vipera ammodytes meridionalis* BOULENGER, 1903
map p. 92 Sikinos, Greece
65 cm
photo: A. MEYER

RS10502-4 *Vipera ammodytes meridionalis* BOULENGER, 1903
map p. 92 Ios Island, Cyclades, Greece
65 cm
male
photo: P. BRODMANN

RS11100-4 *Vipera ammodytes montandoni* BOULENGER, 1904
map p. 92 Strandza National Park, Bulgaria
89 cm
female
photo: D. JABLONSKI

RS11101-4 *Vipera ammodytes montandoni* BOULENGER, 1904
map p. 92 near Babadag, Tulcea, Romania
89 cm
photo: J. VAN DER VOORT

RS11102-4 *Vipera ammodytes montandoni* BOULENGER, 1904
map p. 92 near Babadag, Tulcea, Romania
89 cm
photo: J. VAN DER VOORT

RS11103-4 *Vipera ammodytes montandoni* BOULENGER, 1904
map p. 92 Turkey (European part)
89 cm
photo: F. SIEGENTHALER

RS11110-4 *Vipera ammodytes transcaucasiana* BOULENGER, 1913
map p. 92 Kura Valley, Georgia
75 cm
male
photo: I. PERANIC

RS11111-4 *Vipera ammodytes transcaucasiana* BOULENGER, 1913
map p. 92 Georgia
75 cm
male
photo: D. JABLONSKI

RS11112-3 *Vipera ammodytes transcaucasiana* BOULENGER, 1913
map p. 92 Georgia
75 cm
subadult female
photo: D. JANDZIK

RS11113 Habitat of *Vipera ammodytes transcaucasiana*
Ordu province, Turkey
photo: D. Jandzik

RS11114-2 *Vipera ammodytes transcaucasiana* Boulenger, 1913
map p. 92
Ordu province, Turkey
75 cm
juveniles
photo: D. Jandzik

RS11115-4 *Vipera ammodytes transcaucasiana* Boulenger, 1913
map p. 92
Ordu province, Turkey
75 cm
male
photo: D. Jandzik

RS11116-4 *Vipera ammodytes transcaucasiana* Boulenger, 1913
map p. 92
Ordu province, Turkey
75 cm
gravid female
photo: D. Jandzik

RS11117-4 *Vipera ammodytes transcaucasiana* Boulenger, 1913
map p. 92
Turkey
75 cm
male
photo: D. Jandzik

RS11118-4 *Vipera ammodytes transcaucasiana* Boulenger, 1913
map p. 92
Kose Dag, Turkey
75 cm
female
photo: A. Teynié

RS11120 Habitat of *Vipera anatolica*
Turkey

photo: H. Sigg

RS11121-4 *Vipera anatolica* Eiselt & Baran, 1970
map p. 116 Kohu Dag, Ciglikara Ormanlari, Turkey
43 cm

photo: H. Sigg

RS11130-4 *Vipera aspis aspis* (Linnaeus, 1758)
Southern Black Forest, Germany
82 cm

photo: I. Peranic

RS11131-4 *Vipera aspis aspis* (Linnaeus, 1758)
map p. 105 Germany
82 cm
females

photo: T. Ott

RS11133 Habitat of *Vipera aspis aspis*
Southern Black Forest, Germany

photo: T. Ott

RS11134-4 *Vipera aspis aspis* (Linnaeus, 1758)
map p. 105 Germany
82 cm
male

photo: T. Ott

RS11135 Habitat of *Vipera aspis aspis*
Northwestern Switzerland

photo: T. Ott

RS11136-4 *Vipera aspis aspis* (Linnaeus, 1758)
map p. 105 Fin Dessus, Switzerland
82 cm
female

photo: S. Dummermuth

RS11137-4 *Vipera aspis aspis* (Linnaeus, 1758)
map p. 105 Jura, Switzerland
82 cm
male

photo: T. Ott

RS11138-4 *Vipera aspis aspis* (Linnaeus, 1758)
map p. 105 western Switzerland
82 cm
albinotic specimen

photo: H. Billing

RS11139-4 *Vipera aspis aspis* (LINNAEUS, 1758)
map p. 105 Brüggligrat, Switzerland
82 cm
male
photo: S. DUMMERMUTH

RS11140-4 *Vipera aspis aspis* (LINNAEUS, 1758)
map p. 105 Jura, Switzerland
82 cm
female
photo: T. OTT

RS11141-4 *Vipera aspis aspis* (LINNAEUS, 1758)
map p. 105 Jura, Switzerland
82 cm
female
photo: T. OTT

RS11142-4 *Vipera aspis aspis* (LINNAEUS, 1758)
map p. 105 Jura, Switzerland
82 cm
male
photo: T. OTT

RS11143 Habitat of *Vipera aspis aspis*
France

photo: T. Ott

RS11144-4 *Vipera aspis aspis* (Linnaeus, 1758)
Vaufrey, Doubs Department, France
82 cm
female

photo: S. Dummermuth

RS11145-4 *Vipera aspis aspis* (Linnaeus, 1758)
Limousin, France
82 cm
male

photo: H. van der Ploeg & C. Vermeer

RS11146-4 *Vipera aspis aspis* (Linnaeus, 1758)
Central France
82 cm
female

photo: A. Meyer

RS11147-4 *Vipera aspis aspis* (Linnaeus, 1758)
vicinity of Verdun, Meuse Department, France
82 cm

photo: J. van der Voort

RS11148-4 *Vipera aspis aspis* (Linnaeus, 1758)
vicinity of Verdun, Meuse Department, France
82 cm

photo: J. van der Voort

RS11149-4 *Vipera aspis aspis* (LINNAEUS, 1758)
Saint-Saturnin, Puy de Dôme Department, France
82 cm
male
photo: A. TEYNIÉ

RS11150-4 *Vipera aspis aspis* (LINNAEUS, 1758)
Saint-Saturnin, Puy de Dôme Department, France
82 cm
male
photo: A. TEYNIÉ

RS11151-4 *Vipera aspis aspis* (LINNAEUS, 1758)
Burgos, Spain
82 cm
photo: D. BOHLE

RS11152-4 *Vipera aspis aspis* (LINNAEUS, 1758)
Castilla y León, Spain
82 cm
female
photo: A. MEYER

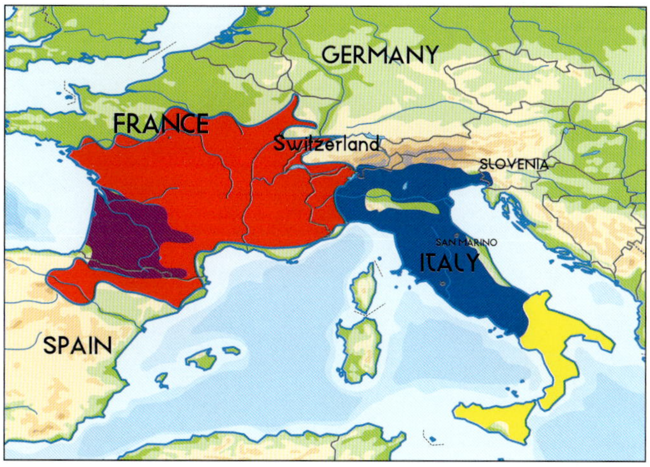

Vipera aspis aspis (red)
Vipera aspis francisciredi (blue)
Vipera aspis hugy (yellow)
Vipera aspis zinnikeri (purple)

RS11153-4 Hybrid of *Vipera aspis* and *Vipera latastei*
photo: B. TRAPP

RS11154-4 *Vipera aspis aspis* (LINNAEUS, 1758) (form *atra*)
map p. 105 Northern Tessine, Switzerland
82 cm
female
photo: T. OTT

RS11155-4 *Vipera aspis aspis* (LINNAEUS, 1758) (form *atra*)
map p. 105 Northern Tessine, Switzerland
82 cm
female
photo: T. OTT

RS11156-4 *Vipera aspis aspis* (LINNAEUS, 1758) (form *atra*)
map p. 105 Northern Tessine, Switzerland
82 cm
female
photo: T. OTT

RS11157-4 *Vipera aspis aspis* (LINNAEUS, 1758) (form *atra*)
map p. 105 Northern Tessine, Switzerland
82 cm
male
photo: T. OTT

RS11158-4 *Vipera aspis aspis* (LINNAEUS, 1758) (form *atra*)
map p. 105 Northern Tessine, Switzerland
82 cm
male
photo: T. OTT

RS11159-4 *Vipera aspis aspis* (LINNAEUS, 1758) (form *atra*)
map p. 105 Northern Tessine, Switzerland
82 cm
male
photo: T. OTT

RS11160-4 *Vipera aspis aspis* (LINNAEUS, 1758) (form *atra*)
map p. 105 Wallis, Switzerland
82 cm
photo: T. OTT

RS11161-4 *Vipera aspis aspis* (LINNAEUS, 1758) (form *atra*)
map p. 105 Northern Tessine, Switzerland
82 cm
male
photo: T. OTT

RS11162-4 *Vipera aspis aspis* (LINNAEUS, 1758) (form *atra*) photo: K. MEBERT
map p. 105 Bern Canton, Switzerland
82 cm

RS11163-4 *Vipera aspis aspis* (LINNAEUS, 1758) (form *atra*)
map p. 105 Tessine Canton, Switzerland
82 cm
photo: K. MEBERT

RS11164-4 *Vipera aspis aspis* (LINNAEUS, 1758) (form *atra*)
map p. 105 Tessine Canton, Switzerland
82 cm
photo: K. MEBERT

RS11165 Habitat of *Vipera aspis aspis* (form *atra*)
Southwestern Switzerland

photo: K. Mebert

RS11166-4 *Vipera aspis aspis* (Linnaeus, 1758) (form *atra*)
map p. 105 Tessine Canton, Switzerland
82 cm

photo: K. Mebert

F

RS11167-4 *Vipera aspis aspis* (Linnaeus, 1758) (form *atra*)
map p. 105 Innertkirchen, Switzerland
82 cm
two females, one melanistic

photo: P. Brodmann

F

RS11168-4 *Vipera aspis aspis* (Linnaeus, 1758) (form *atra*)
map p. 105 Devero, Italy
82 cm
male

photo: H. van der Ploeg & C. Vermeer

F

RS11510-4 *Vipera aspis francisciredi* Laurenti, 1768
map p. 105 Southern Tessine, Switzerland
94 cm
female

photo: T. Ott

D

RS11511 Habitat of *Vipera aspis francisciredi*
Southern Tessine, Switzerland

photo: T. Ott

RS11512-4 *Vipera aspis francisciredi* LAURENTI, 1768
map p. 105 Southern Tessine, Switzerland
94 cm
female
photo: T. OTT

RS11513-4 *Vipera aspis francisciredi* LAURENTI, 1768
map p. 105 Southern Tessine, Switzerland
94 cm
female
photo: T. OTT

RS11514-4 *Vipera aspis francisciredi* LAURENTI, 1768
map p. 105 Southern Tessine, Switzerland
94 cm
female
photo: T. OTT

RS11515-4 *Vipera aspis francisciredi* LAURENTI, 1768
map p. 105 Southern Tessine, Switzerland
94 cm
male
photo: T. OTT

RS11516-4 *Vipera aspis francisciredi* LAURENTI, 1768
map p. 105 Abruzzo Region, Italy
94 cm

photo: D. BOHLE

RS11517 Habitat of *Vipera aspis francisciredi*
Abruzzo Region, Italy

photo: D. BOHLE

RS11518-4 *Vipera aspis francisciredi* LAURENTI, 1768
map p. 105 Abruzzo Region, Italy
94 cm

photo: D. BOHLE

RS11519-4 *Vipera aspis francisciredi* Laurenti, 1768
map p. 105 North of Venezia, Italy
94 cm
male
photo: H. van der Ploeg & C. Vermeer

RS11520-3 *Vipera aspis francisciredi* Laurenti, 1768
map p. 105 Elba Island, Italy
94 cm
subadult female
photo: A. Pieh

RS11710 Habitat of *Vipera aspis hugyi*
Sicilia, Italy
photo: A. Meyer

RS11711-4 *Vipera aspis hugyi* Schinz, 1834
map p. 105 Selimunte, sea level, Sicilia, Italy
67 cm
male
photo: P. Brodmann

RS11712-4 *Vipera aspis hugyi* Schinz, 1834
map p. 105 Calabria, Italy
67 cm
female
photo: I. Peranic

RS11713-4 *Vipera aspis hugyi* Schinz, 1834
map p. 105 Basilicata, Italy
67 cm
female
photo: A. Meyer

RS11714-4 map p. 105
Vipera aspis hugyi SCHINZ, 1834
Basilicata, Italy
67 cm
female
photo: A. MEYER

R11715-4 map p. 105
Vipera aspis hugyi SCHINZ, 1834
Sicilia, Italy
67 cm
male
photo: A. MEYER

RS11716-4 map p. 105
Vipera aspis hugyi SCHINZ, 1834 (form *montecristi*)
Montecristo Island, Italy
67 cm
female
photo: P. BRODMANN

RS11717-4 map p. 105
Vipera aspis hugyi SCHINZ, 1834 (form *montecristi*)
Montecristo Island, Italy
67 cm
male
photo: P. BRODMANN

RS11718-4 map p. 105
Vipera aspis hugyi SCHINZ, 1834 (form *montecristi*)
Montecristo Island, Italy
67 cm
male
photo: P. BRODMANN

RS11719-4 map p. 105
Vipera aspis hugyi SCHINZ, 1834 (form *montecristi*)
Montecristo Island, Italy
67 cm
photo: J. TASHJIAN

RS11720-4 *Vipera aspis zinnikeri* KRAMER, 1958
map p. 105 Pyrénées Orientales Department, France
71 cm
female
photo: A. MEYER

RS11721-4 *Vipera aspis zinnikeri* KRAMER, 1958
map p. 105 Hautes Pyrénées Department, France
71 cm
female
photo: D. BOHLE

RS11722-4 *Vipera aspis zinnikeri* KRAMER, 1958
map p. 105 Pyrénées Mts., France
71 cm
male, mountain form
photo: P. JANSSEN

RS11723-4 *Vipera aspis zinnikeri* Kramer, 1958
map p. 105 Hautes Pyrénées Department, France
71 cm
female
photo: A. Meyer

RS11724-4 *Vipera aspis zinnikeri* Kramer, 1958
map p. 105 Hautes Pyrénées Department, France
71 cm
female
photo: A. Meyer

RS11725-4 *Vipera aspis zinnikeri* Kramer, 1958
map p. 105 Soulac, Gironde Department, France
71 cm
male
photo: A. Teynié

RS11726-4 *Vipera aspis zinnikeri* Kramer, 1958
map p. 105 Soulac, Gironde Department, France
71 cm
male
photo: A. Teynié

RS11727-2 *Vipera aspis zinnikeri* Kramer, 1958
map p. 105 Toulouse, France
71 cm
juvenile
photo: J. Maran

RS11728-4 *Vipera aspis zinnikeri* Kramer, 1958
map p. 105 Toulouse, France
71 cm
female
photo: J. Maran

RS11730 Habitat of *Vipera barani*
Ordu Province, Turkey

photo: P. Geniez & A. Teynié

Vipera anatolica (red)
Vipera barani (blue)

RS11731-4 *Vipera barani* Böhme & Joger, 1984
Ordu Province, Turkey
59 cm

photo: A. Avci

RS11732-2 *Vipera barani* Böhme & Joger, 1984
Ordu Province, Turkey
59 cm
juvenile

photo: P. Geniez & A. Teynié

RS11733-4 *Vipera barani* Böhme & Joger, 1984
Turkey
59 cm

photo: F. Siegenthaler

RS11734-4 *Vipera barani* Böhme & Joger, 1984
Ordu Province, Turkey
59 cm
female

photo: P. Geniez & A. Teynié

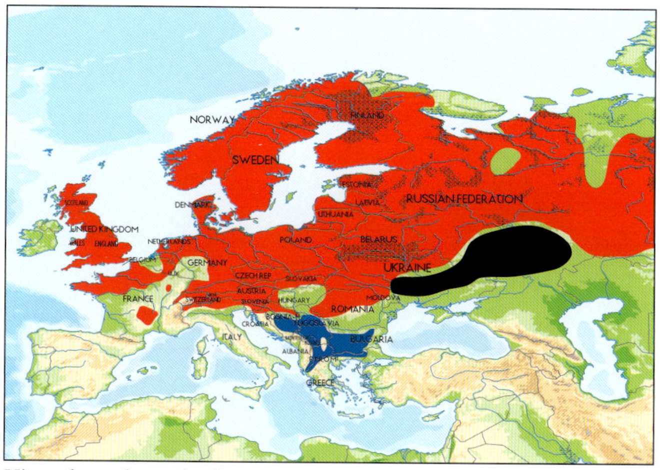

Vipera berus berus (red)
Vipera berus bosniensis (blue)
Vipera berus nikolskii (black)

Vipera berus berus (red)
Vipera berus nikolskii (black)
Vipera berus sachalinensis (green)

RS11740-4 *Vipera berus berus* (LINNAEUS, 1758)
Göta Canal, Linköping, Sweden
94 cm
female
photo: M. KARLSSON

RS11741-4 *Vipera berus berus* (LINNAEUS, 1758)
State of Mecklenburg-Vorpommern, Germany
94 cm
female
photo: D. BOHLE

RS11742-4 *Vipera berus berus* (LINNAEUS, 1758)
State of Brandenburg, Germany
94 cm
photo: D. BOHLE

RS11743 Habitat of *Vipera berus berus*
State of Brandenburg, Germany
photo: D. BOHLE

RS11744-4 *Vipera berus berus* (LINNAEUS, 1758)
map p. 117 State of Brandenburg, Germany
94 cm
male combat
photo: D. BOHLE

RS11745-4 *Vipera berus berus* (LINNAEUS, 1758)
map p. 117 State of Hessen, Germany
94 cm
male
photo: D. BOHLE

RS11746-4 *Vipera berus berus* (LINNAEUS, 1758)
map p. 117 Spessart, Germany
94 cm
male
photo: J.M. DEHLING

RS11747-4 *Vipera berus berus* (LINNAEUS, 1758)
map p. 117 State of Thüringen, Germany
94 cm
male
photo: D. BOHLE

RS11748-4 *Vipera berus berus* (LINNAEUS, 1758)
map p. 117 Samedan, 1850 m a.s.l., State of Bavaria, Germany
94 cm
melanistic female, giving birth
photo: P. BRODMANN

RS11749-4 map p. 117 *Vipera berus berus* (LINNAEUS, 1758)
Northern Black Forest, Germany
94 cm
female, melanistic form
photo: J.M. DEHLING

RS11750 Habitat of *Vipera berus berus* (LINNAEUS, 1758)
Black Forest, Germany
photo: T. OTT

RS11751-2 map p. 117 *Vipera berus berus* (LINNAEUS, 1758)
Drenthe, Netherlands
94 cm
juvenile
photo: H. VAN DER PLOEG

RS11752-4 map p. 117 *Vipera berus berus* (LINNAEUS, 1758)
Drenthe, Netherlands
94 cm
photo: H. VAN DER PLOEG

RS11753-4 map p. 117 *Vipera berus berus* (LINNAEUS, 1758)
Drenthe, Netherlands
94 cm
female
photo: H. VAN DER PLOEG

RS11754-2 map p. 117 *Vipera berus berus* (LINNAEUS, 1758)
Lac du Guéry, Orcival, Puy-de-Dôme Department, France – 94 cm
juvenile female
photo: A. TEYNIÉ

RS11755-2 *Vipera berus berus* (LINNAEUS, 1758)
map p. 117 Lac de Montcynère, Compains, Puy-de-Dôme Department, France – 94 cm
juvenile female
photo: A. TEYNIÉ

RS11756-4 *Vipera berus berus* (LINNAEUS, 1758)
map p. 117 Jeseniky Mts., Czech Republic
94 cm
female
photo: V. GVOZDIK

RS11757 Habitat of *Vipera berus berus*
Nizke Tatry National Park, Slovakia
photo: D. JABLONSKI

RS11758-4 *Vipera berus berus* (LINNAEUS, 1758)
map p. 117 Mala Fatra National Park, Slovakia
94 cm
female (morph *chersea*)
photo: D. JABLONSKI

RS11759-4 *Vipera berus berus* (LINNAEUS, 1758)
map p. 117 Zlar, Slovakia
94 cm
female
photo: D. JANDZIK

RS11760-4 *Vipera berus berus* (LINNAEUS, 1758)
map p. 117 Tessine Canton, Switzerland
94 cm
male
photo: T. OTT

RS11761-4 *Vipera berus berus* (LINNAEUS, 1758) photo: K. MEBERT
map p. 117 Uri Canton, Switzerland
94 cm

RS11762-4 *Vipera berus berus* (LINNAEUS, 1758)
map p. 117 Southern Tirol, Italy
94 cm
male, unusual pattern photo: K. PIEH

RS11763-4 *Vipera berus berus* (LINNAEUS, 1758)
map p. 117 Barnova Woods, Iasi, Romania
94 cm
photo: J. VAN DER VOORT

RS11764-4 *Vipera berus berus* (LINNAEUS, 1758)
map p. 117 Northern Carpathians, Romania
94 cm
male
photo: I. PERANIC

RS11765-3 *Vipera berus berus* (LINNAEUS, 1758)
map p. 117 Gorski Kotar, Croatia
94 cm
subadult female
photo: I. PERANIC

RS11766-4 *Vipera berus berus* (LINNAEUS, 1758) × *V. berus nikolskii* VEDMEDERJA, GRUBANT & RUDAYEVA, 1986
map p. 117 Confluence of Kama river and Volga river, Tatarstan Republic, Russia
photo: A. PAVLOV

RS11767-4 *Vipera berus berus* (LINNAEUS, 1758) × *V. berus nikolskii* VEDMEDERJA, GRUBANT & RUDAYEVA, 1986
map p. 117 Tatarstan Republic, Russia
female with mixed characteristics
photo: A. PAVLOV

RS11768 Habitat of *Vipera berus berus* × *V. berus nikolskii*
Tatarstan Republic, Russia
photo: A. PAVLOV

RS12000-4 *Vipera berus bosniensis* BOETTGER *in* MOJSISOVICS, 1889 – Montenegro
map p. 117
67 cm
female
photo: T. OTT

RS12001-4 *Vipera berus bosniensis* BOETTGER *in* MOJSISOVICS, 1889 – Montenegro
map p. 117
67 cm
male
photo: T. OTT

RS12002-4 *Vipera berus bosniensis* BOETTGER *in* MOJSISOVICS, 1889 – Kolašin, Montenegro
map p. 117
67 cm
female
photo: D. BOHLE

RS12003 Habitat of *Vipera berus bosniensis*
Kolašin, Montenegro
photo: D. BOHLE

RS12004-4 *Vipera berus bosniensis* BOETTGER *in* MOJSISOVICS, 1889 – Mts. of Southwestern Bosnia
map p. 117
67 cm
male
photo: I. PERANIC

RS12005-4 map p. 117 *Vipera berus bosniensis* BOETTGER *in* MOJSISOVICS, 1889 – Makedonia, Greece
67 cm
male
photo: A. MEYER

RS12006-4 map p. 117 *Vipera berus bosniensis* BOETTGER *in* MOJSISOVICS, 1889 – Rila Mts., Bulgaria
67 cm
photo: D. JANDZIK

RS12007-4 map p. 117 *Vipera berus bosniensis* BOETTGER *in* MOJSISOVICS, 1889 – Mts. of Southwestern Bosnia
67 cm
melanistic male
photo: I. PERANIC

RS12008-4 map p. 117 *Vipera berus bosniensis* BOETTGER *in* MOJSISOVICS, 1889 – Central Bosnia
67 cm
male, highland morph
photo: I. PERANIC

RS12009-2 map p. 117 *Vipera berus bosniensis* BOETTGER *in* MOJSISOVICS, 1889 – Surroundings of Zagreb, Croatia
67 cm
juvenile female
photo: D. JABLONSKI

RS12010-2 map p. 117 *Vipera berus bosniensis* BOETTGER *in* MOJSISOVICS, 1889 – Zagreb, Croatia
67 cm
juvenile female
photo: I. PERANIC

RS12011-4 *Vipera berus bosniensis* BOETTGER *in* MOJSISOVICS, 1889 – Zagreb, Croatia
map p. 117
67 cm
female
photo: I. PERANIC

RS12012-4 *Vipera berus bosniensis* BOETTGER *in* MOJSISOVICS, 1889 – Zagreb, Croatia
map p. 117
67 cm
female
photo: I. PERANIC

RS12013-4 *Vipera berus bosniensis* BOETTGER *in* MOJSISOVICS, 1889
map p. 117
Turopolje, southeast of Zagreb, Croatia
67 cm
male, lowland form
photo: B. HORVATIC

RS12200-2 *Vipera berus nikolskii* Vedmederja, Grubant & Rudayeva, 1986 – Kharkov, Ukraine
map p. 117
87 cm
juvenile female
photo: D. Jablonski

RS12201-2 *Vipera berus nikolskii* Vedmederja, Grubant & Rudayeva, 1986 – Kharkov, Ukraine
map p. 117
87 cm
juvenile female
photo: D. Jablonski

RS12202-2 *Vipera berus nikolskii* Vedmederja, Grubant & Rudayeva, 1986 – Ukraine
map p. 117
87 cm – juvenile with congenital kyphoscoliosis
photo: D. Jandzik

RS12203-4 *Vipera berus nikolskii* Vedmederja, Grubant & Rudayeva, 1986 – Ukraine
map p. 117
87 cm
photo: D. Jandzik

RS12204-4 *Vipera berus nikolskii* Vedmederja, Grubant & Rudayeva, 1986 – Zubrilovo, Penza Region, Russia
map p. 117
87 cm
female
photo: H. van der Ploeg

RS12205-4 *Vipera berus nikolskii* Vedmederja, Grubant & Rudayeva, 1986 – Kharkov, Ukraine
map p. 117
87 cm
photo: D. Jablonski

RS12210-4 *Vipera berus sachalinensis* TSAREWSKY, 1916
map p. 117 Russia
78 cm
female
photo: D. JANDZIK

RS12211-4 *Vipera berus sachalinensis* TSAREWSKY, 1916
map p. 117 Russia
78 cm
female
photo: L. TRUTNAU

RS12220 Habitat of *Vipera darevskii*
Artwin Province, Turkey
photo: P. GENIEZ & A. TEYNIÉ

RS12221-2 *Vipera darevskii* VEDMEDERJA, ORLOV & TUNIYEV,
map p. 133 1986 – Artwin Province, Turkey
42 cm
juvenile
photo: P. GENIEZ & A. TEYNIÉ

RS12222-2 *Vipera darevskii* VEDMEDERJA, ORLOV & TUNIYEV,
map p. 133 1986 – Artwin Province, Turkey
42 cm
juvenile male
photo: P. GENIEZ & A. TEYNIÉ

RS12223-4 *Vipera darevskii* VEDMEDERJA, ORLOV & TUNIYEV,
map p. 133 1986 – Javakheti Mts., Armenia
42 cm
male
photo: I. PERANIC

RS12230 Habitat of *Vipera dinniki*
Northern Caucasus, Republic of Adygea, Russia
photo: L. Darafejenka

RS12231-4 *Vipera dinniki* Nikolsky, 1913
map p. 133 Northern Caucasus, Republic of Adygea, Russia
59 cm
photo: L. Darafejenka

RS12232-4 *Vipera dinniki* Nikolsky, 1913
map p. 133 Caucasus
59 cm
female
photo: L. Westrin

RS12233-4 *Vipera dinniki* NIKOLSKY, 1913
map p. 133
Northern Caucasus, Republic of Adygea, Russia
59 cm
female

photo: L. DARAFEJENKA

RS12234-4 *Vipera dinniki* NIKOLSKY, 1913
map p. 133
Caucasus
59 cm

photo: B. TRAPP

RS12235-4 *Vipera dinniki* NIKOLSKY, 1913
map p. 133
Caucasus
59 cm

photo: B. TRAPP

RS12240 Habitat of *Vipera ebneri* and *Gloydius halys caucasicus*
Laar Valley, Alborz (or Elburz) Mts., Iran

photo: B. Trapp

RS12241-4 *Vipera ebneri* Knoepffler & Sochurek, 1955
map p. 138 Laar Valley, Alborz (or Elburz) Mts., Iran
44 cm

photo: B. Trapp

RS12242-4 *Vipera ebneri* Knoepffler & Sochurek, 1955
map p. 138 Laar Valley, Alborz (or Elburz) Mts., Iran
44 cm

photo: B. Trapp

RS12243-4 *Vipera ebneri* Knoepffler & Sochurek, 1955
map p. 138 Laar Valley, Alborz (or Elburz) Mts., Iran
44 cm

photo: B. Trapp

RS12244-4 *Vipera ebneri* Knoepffler & Sochurek, 1955
map p. 138 Serum Institute, Teheran, Iran
44 cm

photo: N. Beigi

RS12250-4 *Vipera eriwanensis* (Reuss, 1933)
map p. 138 Kars Province, Turkey
51 cm
female
photo: D. Jandzik

RS12251 Habitat of *Vipera eriwanensis*
Kars Province, Turkey
photo: D. Jandzik

RS12252-4 *Vipera eriwanensis* (Reuss, 1933)
map p. 138 Kars Province, Turkey
51 cm
male
photo: D. Jandzik

RS12253-4 *Vipera eriwanensis* (Reuss, 1933)
map p. 138 Sewán Lake, Yerevan, Armenia
51 cm
male
photo: P. Brodmann

RS12254-4 *Vipera eriwanensis* (Reuss, 1933)
map p. 138 Kars Province, Turkey
51 cm
female
photo: P. Geniez & A. Teynié

RS12255-4 *Vipera eriwanensis* (Reuss, 1933)
map p. 138 Kars Province, Turkey
51 cm
male
photo: A. Teynié

RS12260 Habitat of *Vipera kaznakovi*
region of Batumi, Southern Autonomous Republic of Adjara, Georgia
photo: T. BADER

RS12261-4 *Vipera kaznakovi* NIKOLSKY, 1909
region of Batumi, Southern Autonomous Republic of Adjara, Georgia – 65 cm
male
photo: T. BADER

RS12262-4 *Vipera kaznakovi* NIKOLSKY, 1909
region of Batumi, Southern Autonomous Republic of Adjara, Georgia
65 cm
male
photo: T. BADER

Vipera kaznakovi (red)
Vipera darevskii (purple)
Vipera dinniki (blue)

RS12263-4 *Vipera kaznakovi* NIKOLSKY, 1909
Hopa, Artvin Province, Turkey
65 cm
female
photo: H. VAN DER PLOEG & C. VERMEER

RS12264-4 *Vipera kaznakovi* NIKOLSKY, 1909
Hopa, Artvin Province, Turkey
65 cm
female
photo: D. JANDZIK

RS12265-4 *Vipera kaznakovi* NIKOLSKY, 1909
Hopa, Artvin Province, Turkey
65 cm
male
photo: P. GENIEZ & A. TEYNIÉ

RS12266-4 *Vipera kaznakovi* NIKOLSKY, 1909
Hopa, Artvin Province, Turkey
65 cm
male
photo: H. VAN DER PLOEG & C. VERMEER

RS12267-2 *Vipera kaznakovi* NIKOLSKY, 1909
Hopa, Artvin Province, Turkey
65 cm
juvenile female
photo: A. TEYNIÉ

RS12270 Habitat of *Vipera latastei latastei*
region of Castilla y León, Spain

photo: A. Meyer

RS12271-4 *Vipera latastei latastei* Boscá, 1878
region of Castilla y León, Spain
72 cm
female

photo: A. Meyer

RS12272-4 *Vipera latastei latastei* Boscá, 1878
Province of Castellón, Spain
72 cm

photo: K. Mebert

RS12273-4 *Vipera latastei latastei* Boscá, 1878
Spain
72 cm
male
photo: D. Jablonski

RS12274-4 *Vipera latastei latastei* Boscá, 1878
Spain
72 cm
female
photo: D. Jablonski

RS12275-4 *Vipera latastei latastei* Boscá, 1878
Burgos, Spain
72 cm
female
photo: H. van der Ploeg & C. Vermeer

RS12276-4 *Vipera latastei latastei* Boscá, 1878
Burgos, Spain
72 cm
male
photo: A. Teynié

RS12277-4 *Vipera latastei latastei* Boscá, 1878
Burgos, Spain
72 cm
male
photo: P. Janssen

RS12278-4 *Vipera latastei latastei* Boscá, 1878
Serra do Geres, Portugal
72 cm
photo: J. van der Voort

RS12280 Habitat of *Vipera latastei gaditana*
Vivora Retin, Spain

photo: D. Donaire

Vipera latastei latastei (red)
Vipera latastei gaditana (blue)
Vipera latastei cf. *gaditana* (purple)

RS12281-4 *Vipera latastei gaditana* Saint Girons, 1977
Cota Doñana, Region of Andalusia, Spain
60 cm

photo: J. van der Voort

RS12282-4 *Vipera latastei gaditana* Saint Girons, 1977
Cota Doñana, Region of Andalusia, Spain
60 cm

photo: M. Dobiey

RS12283-4 *Vipera latastei gaditana* Saint Girons, 1977
Province of Huelva, Spain
60 cm
male

photo: I. Peranic

RS12284-2 *Vipera latastei gaditana* Saint Girons, 1977
Mazagon, Province of Huelva, Spain
60 cm
juvenile

photo: D. Donaire

RS12285-4 *Vipera latastei gaditana* Saint Girons, 1977
Cota Doñana, Region of Andalusia, Spain
60 cm

photo: J. van der Voort

RS12286-4 *Vipera latastei gaditana* Saint Girons, 1977
Cota Doñana, Region of Andalusia, Spain
60 cm
female

photo: M. Dobiey

RS12287-4 *Vipera latastei* cf. *gaditana* Saint Girons, 1977
Tunisia
60 cm
female

photo: I. Peranic

RS12290-4 *Vipera lotievi* Nilson, Tuniyev, Orlov, Höggren & Andrén, 1995 – Caucasus
60 cm
female
photo: O. Klingner

RS12291-4 *Vipera lotievi* Nilson, Tuniyev, Orlov, Höggren & Andrén, 1995 – Caucasus
60 cm
male
photo: O. Klingner

Vipera ebneri (purple)
Vipera eriwanensis (blue)
Vipera lotievi (red)

RS12292-4 *Vipera lotievi* Nilson, Tuniyev, Orlov, Höggren & Andrén, 1995 – Caucasus, Russia
60 cm
females
photo: I. Peranic

RS12300 Habitat of *Vipera magnifica*
Northern Caucasus, Russia
photo: L. Darafeyenka

RS12301-4 *Vipera magnifica* Tuniyev & Ostrovskikh, 2001
map p. 141
Northern Caucasus, Russia
50 cm
male
photo: L. Darafeyenka

RS12310-4 *Vipera orlovi* Tuniyev & Ostrovskikh, 2001
Russia
52 cm
male
photo: O. Klingner

RS12311-4 *Vipera orlovi* Tuniyev & Ostrovskikh, 2001
Russia
52 cm
female
photo: O. Klingner

RS12312 Habitat of *Vipera orlovi*
Mt. Indjuk, Caucasus, Russia
photo: K. Lipponen

RS12313-4 *Vipera orlovi* TUNIYEV & OSTROVSKIKH, 2001
Western Caucasus, Russia
52 cm
male
photo: M. DOBIEY

RS12314-4 *Vipera orlovi* TUNIYEV & OSTROVSKIKH, 2001
Northwestern Caucasus, Russia
52 cm
female
photo: I. PERANIC

RS12315-4 *Vipera orlovi* TUNIYEV & OSTROVSKIKH, 2001
Northwestern Caucasus, Russia
52 cm
male
photo: I. PERANIC

RS12316-4 *Vipera orlovi* Tuniyev & Ostrovskikh, 2001

52 cm
male and female
photo: L. Westrin

RS12317-2 *Vipera orlovi* Tuniyev & Ostrovskikh, 2001
Western Caucasus, Russia
52 cm
juvenile female
photo: M. Dobiey

Vipera magnifica (blue)
Vipera orlovi (red)
Vipera pontica (purple)

RS12320-4 "*Vipera pontica*" Billing, Nilson & Sattler, 1990
Borcka, Artvin Province, Turkey
30 cm?
photo: H. Billing

RS12321-4 "*Vipera pontica*" Billing, Nilson & Sattler, 1990,
holotype, Borcka, Artvin Province, Turkey
30 cm?
photo: M. Schweiger

RS12322-4 "*Vipera pontica*" Billing, Nilson & Sattler, 1990
Borcka, Artvin Province, Turkey
30 cm?
photo: H. Billing

RS12330 Habitat of *Vipera renardi renardi*
Karagala Steppe, Kazakhstan
photo: A. Pavlov

Vipera renardi renardi (red)
Vipera renardi bashkirovi (purple)
Vipera renardi parursinii, extralimital (yellow)
Vipera renardi tienshanica (blue)

RS12331-4 *Vipera renardi renardi* (Christoph, 1861)
Karagala Steppe, Kazakhstan
62 cm
male
photo: A. Pavlov

RS12332-4 *Vipera renardi renardi* (Christoph, 1861)
border between Romania and Moldova
62 cm

photo: O. Klingner

RS12333-4 *Vipera renardi renardi* (Christoph, 1861)
Eastern Ukraine
62 cm
male

photo: O. Klingner

RS12334-4 *Vipera renardi renardi* (CHRISTOPH, 1861)
map p. 142 Volgodonsk, Russia
62 cm
male
photo: I. PERANIC

RS12335-3 *Vipera renardi renardi* (CHRISTOPH, 1861)
map p. 142 Volgodonsk, Russia
62 cm
subadult female
photo: I. PERANIC

RS12336-4 *Vipera renardi renardi* (CHRISTOPH, 1861)
map p. 142 Estuary of Dnepr River, near Kimburk, Ukraine
62 cm
photo: O. KNOBLOCH

R12337-4 *Vipera renardi renardi* (CHRISTOPH, 1861)
map p. 142 Danube Delta, Romania
62 cm
male
photo: O. KLINGNER

RS12338-4 *Vipera renardi renardi* (CHRISTOPH, 1861)
map p. 142 Near Tiraspol, Moldova
62 cm
photo: D. JABONSKI

RS12339-4 *Vipera renardi renardi* (CHRISTOPH, 1861)
map p. 142 Crimean form, Cape Chauda, Kerch Peninsula, Ukraine – 62 cm
female
photo: M.M. BESKARAVAINYI, courtesy of O.V. KUKUSHKIN

RS12340 Habitat of *Vipera renardi bashkirovi*
Spassk Island, Volga River, Republic of Tatarstan, Russia

photo: A. Pavlov

RS12341-2 *Vipera renardi bashkirovi* Bakiev, Garanin, Litvinov, Pavlov & Ratnikov, 2004 – Spassk Island, Volga River, Republic of Tatarstan, Russia – 71 cm juvenile

map p. 142

photo: A. Pavlov

RS12342-4 *Vipera renardi bashkirovi* Bakiev, Garanin, Litvinov, Pavlov & Ratnikov, 2004
Garanin, Spassk Island, Volga River, Republic of Tatarstan, Russia
71 cm
male

map p. 142

photo: T. Paasikunnas

RS12343-4 *Vipera renardi bashkirovi* Bakiev, Garanin, Litvinov, Pavlov & Ratnikov, 2004 photo: A. Pavlov
map p. 142 small islet of Spassk Archipelago, Volga River, Republic of Tatarstan, Russia
71 cm
female

RS1234-4 *Vipera renardi bashkirovi* Bakiev, Garanin, Litvinov, Pavlov & Ratnikov, 2004 photo: A. Pavlov
map p. 142 Spassk Island, Volga River, Republic of Tatarstan, Russia
71 cm
female

RS12350-4 *Vipera renardi tienshanica* NILSON & ANDRÉN, 2001
map p. 142 Kazahkstan
46 cm
male
photo: D. JABLONSKI

RS12351-2 *Vipera renardi tienshanica* NILSON & ANDRÉN, 2001
map p. 142 Kazahkstan
46 cm
juvenile male
photo: D. JABLONSKI

RS12360 Habitat of *Vipera seoanei seoanei*
Region of Asturias, Spain
photo: A. MEYER

RS12361-4 *Vipera seoanei seoanei* LATASTE, 1879
map p. 151 Region of Asturias, Spain
72 cm
male
photo: A. MEYER

RS12362-4 *Vipera seoanei seoanei* LATASTE, 1879
map p. 151 Region of Asturias, Spain
72 cm
female

photo: A. MEYER

RS12363-4 *Vipera seoanei seoanei* LATASTE, 1879
map p. 151 Pyrénées-Atlantiques Department, France
72 cm
male

photo: A. MEYER

RS12364-4 *Vipera seoanei seoanei* LATASTE, 1879
map p. 151 Pyrénées Atlantiques Department, France
72 cm
male
photo: D. BOHLE

RS12365-4 *Vipera seoanei seoanei* LATASTE, 1879
map p. 151 Province of Cantabria, Spain
72 cm
male
photo: D. BOHLE

RS12366-4 *Vipera seoanei seoanei* LATASTE, 1879
map p. 151 Province of Cantabria, Spain
72 cm
female
photo: D. BOHLE

RS12367-4 *Vipera seoanei seoanei* LATASTE, 1879
Province of Cantabria, Spain
72 cm
female
photo: A. MEYER

RS12368-4 *Vipera seoanei seoanei* LATASTE, 1879
Province of Cantabria, Spain
72 cm
female
photo: A. MEYER

RS12369-4 *Vipera seoanei seoanei* LATASTE, 1879
Province of Cantabria, Spain
72 cm
male
photo: A. MEYER

RS12370-4 *Vipera seoanei seoanei* LATASTE, 1879
North of Santander, Spain
72 cm
male
photo: H. VAN DER PLOEG & C. VERMEER

RS12371-4 *Vipera seoanei seoanei* LATASTE, 1879
Serra da Peneda, Portugal
72 cm
photo: J. VAN DER VOORT

RS12372-4 *Vipera seoanei* LATASTE, 1879, specimen intermediate between *V. seoanei seoanei* and *V. seoanei cantabrica*
Spain
male
photo: P. JANSSEN

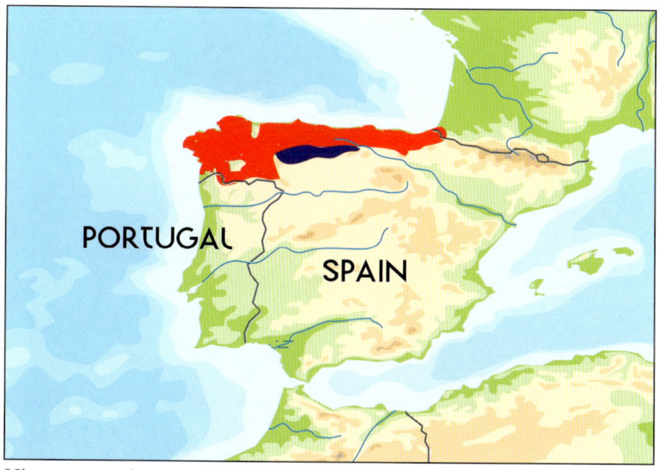

Vipera seoanei seoanei (red)
Vipera seoanei cantabrica (blue)

RS12380 Habitat of *Vipera seoanei cantabrica*
Region of Castilla y León, Spain

photo: A. Meyer

RS12381-4 *Vipera seoanei cantabrica* Braña & Bas, 1983
Region of Castilla y León, Spain
66 cm
female
photo: A. Meyer

RS12382-4 *Vipera seoanei cantabrica* Braña & Bas, 1983
Province of León, Spain
66 cm
female
photo: I. Peranic

RS12383-4 *Vipera seoanei cantabrica* Braña & Bas, 1983
Picos de Europa, Spain
66 cm
male
photo: M. Dobiey

RS12384-4 *Vipera seoanei cantabrica* Braña & Bas, 1983
Southwest of Picos de Europa, Spain
66 cm
male
photo: H. van der Ploeg & C. Vermeer

RS12520-4 *Vipera ursinii ursinii* (Bonaparte, 1835)
Abruzzo Region, Italy
48 cm
photo: D. Bohle

RS12521-4 *Vipera ursinii ursinii* (Bonaparte, 1835)
Abruzzo Region, Italy
48 cm
photo: D. Bohle

RS12522-4 *Vipera ursinii ursinii* (Bonaparte, 1835)
Abruzzo Region, Italy
48 cm
photo: D. Bohle

RS12523-4 *Vipera ursinii ursinii* (BONAPARTE, 1835)
France
48 cm
female
photo: T. OTT

RS12524-4 *Vipera ursinii ursinii* (BONAPARTE, 1835)
France
48 cm
male
photo: T. OTT

RS12525-4 *Vipera ursinii ursinii* (BONAPARTE, 1835)
Abruzzo Region, Italy
48 cm
photo: D. BOHLE

RS12526-4 *Vipera ursinii ursinii* (Bonaparte, 1835)
Central Italy
48 cm
copulation

photo: K. Mebert

RS12527-4 *Vipera ursinii ursinii* (Bonaparte, 1835)
Province of Umbria, Italy
48 cm

photo: D. Bohle

RS12530-4 *Vipera ursinii graeca* NILSON & ANDRÉN, 1988
Pindus Mts., Ioannina, Greece
44 cm

photo: J. VAN DER VOORT

RS12531 Habitat of *Vipera ursinii graeca*
Pindus Mts., Greece

photo: B. TRAPP

RS12532-4 *Vipera ursinii graeca* NILSON & ANDRÉN, 1988
Pindus Mts., Greece
44 cm

photo: B. TRAPP

Vipera ursinii ursinii (red)
Vipera ursinii graeca (purple)
Vipera ursinii macrops (blue)
Vipera ursinii moldavica (green)
Vipera ursinii rakosiensis (black)

RS12540 Habitat of *Vipera ursinii macrops*
Central Croatia
photo: D. Jablonski

RS12541-4 *Vipera ursinii macrops* Méhelÿ, 1911
map p. 155
Dinara Mts., Croatia
47 cm
female
photo: I. Peranic

RS12542-4 *Vipera ursinii macrops* Méhelÿ, 1911
map p. 155
Velebit, Croatia
47 cm
male
photo: B. Horvatic

RS12543-4 *Vipera ursinii macrops* MÉHELŸ, 1911
map p. 155 Dinara Mts., Croatia
47 cm
male

photo: I. PERANIC

RS12544-4 *Vipera ursinii macrops* MÉHELŸ, 1911
map p. 155 Zelengora Mts., Bosnia
47 cm
pair

photo: K. MEBERT

RS12545-4 *Vipera ursinii macrops* MÉHELŸ, 1911
map p. 155 Montenegro
47 cm

photo: E. EVEN

RS12550-4 map p. 155 *Vipera ursinii moldavica* Nilson, Andrén & Joger, 1993 – type locality, Iasi, Romania
56 cm
female
photo: O. Klingner

RS12551-4 map p. 155 *Vipera ursinii moldavica* Nilson, Andrén & Joger, 1993 – Vitava, Romania
56 cm
photo: O. Klingner

RS12552-4 map p. 155 *Vipera ursinii moldavica* Nilson, Andrén & Joger, 1993 – Grinduls, Danube Delta, Tulcea, Romania
56 cm
photo: J. van der Voort

RS12553-4 map p. 155 *Vipera ursinii moldavica* Nilson, Andrén & Joger, 1993 – Grinduls, Danube Delta, Tulcea, Romania
56 cm
photo: J. van der Voort

RS12554-4 map p. 155 *Vipera ursinii moldavica* Nilson, Andrén & Joger, 1993 – Tulcea, Romania
56 cm
female
photo: O. Klingner

RS12555-4 map p. 155 *Vipera ursinii moldavica* Nilson, Andrén & Joger, 1993 – Tulcea, Romania
56 cm
male
photo: O. Klingner

RS12556-4 map p. 155 *Vipera ursinii moldavica* Nilson, Andrén & Joger, 1993 – Grinduls, Danube Delta, Tulcea, Romania 56 cm

photo: J. van der Voort

RS12560 Habitat of *Vipera ursinii rakosiensis* Klausenburg, Romania

photo: O. Klingner

RS12561-4 map p. 155 *Vipera ursinii rakosiensis* Méhelÿ, 1894 Kiskunsag, Central Hungary 64 cm

photo: M. Dobiey

RS12562-4 *Vipera ursinii rakosiensis* MÉHELY̌, 1894
map p. 155 Southern Hungary
64 cm
female
photo: G. KREINER

RS12563-4 *Vipera ursinii rakosiensis* MÉHELY̌, 1894
map p. 155 Kumpeszér, Hungary
64 cm
male
photo: P. BRODMANN

RS12564-2 *Vipera ursinii rakosiensis* MÉHELY̌, 1894
map p. 155 Romania
64 cm
juvenile male
photo: O. KLINGNER

NORWAY
ICELAND
DENMARK
UK
SCOTLAND
ULSTER
UNITED
ENGLAND
IRELAND
KINGDOM
WALES
NORWAY
SWEDEN
DENMARK
NETHERLANDS
BEL.
GERMANY
POLAND
CZECH REPUBLIC
SLOVAKIA
FRANCE
SWITZ.
LIECHT.
AUSTRIA
SLOVENIA
HUNGARY
ITALY
MONACO
SAN MARINO
VATICAN CITY
CROATIA
BOSNIA & H.
YUGOSLAVIA
MONTENEGRO
KOSOVO
ALBANIA
ANDORRA
PORTUGAL
SPAIN
PORT.
UK
SPAIN
MALTA